MAKING GENETICS AND GENOMICS POLICY IN BRITAIN

This important book traces the history of genetics and genomics policy in Britain. Detailing the scientific, political, and economic factors that have informed policy and the development of new health services, the book highlights the particular importance of the field of Public Health Genomics.

Although focused primarily on events in Britain, the book reveals a number of globally applicable lessons. The authors explain how and why Public Health Genomics developed and the ways in which genetics and genomics have come to have a central place in many important health debates. Consideration of their ethical, social, and legal implications and ensuring that new services are equitable, appropriate, and well-targeted will be central to effective health planning and policymaking in future.

The book features:

- Interviews with leading individuals who were intimately involved in the development of genetics and genomics policy and Public Health Genomics.
- Insights from experts who participated in a pair of 'witness seminars'.
- Historical analysis exploiting a wide range of primary sources.

Written in a clear and accessible style, this book will be of interest to those involved in the research and practice of genetics, genomics, bioethics, and population health, but also to NHS staff, policymakers, politicians, and the public. It will also be valuable supplementary reading for students of the History of Medicine and Health, Public Health, and Biomedical Sciences.

Philip Begley is a historian of contemporary Britain, with particular interests in health, politics, and public policy. He has been a member of the 'Governance of

Health' team at the University of Liverpool since 2015 and is currently the lead researcher on a project which investigates the history of genetics and genomics policy in Britain. His previous research focused on the history of management in the National Health Service and the emergence of management consultants as important players in health policymaking.

Sally Sheard is Executive Dean of the Institute of Population Health at the University of Liverpool. She also holds the Andrew Geddes and John Rankin Chair of Modern History. Her primary research interest is in the interface between expert advisers and health policymakers. Sally has extensive experience of using history in public and policy engagement. She has also written for and presented television and radio programmes, including the 2018 BBC Radio 4 series *National Health Stories*.

MAKING GENETICS AND GENOMICS POLICY IN BRITAIN

From Personal to Population Health

Philip Begley and Sally Sheard

LONDON AND NEW YORK

Cover image: Getty images

First published 2023
by Routledge
605 Third Avenue, New York, NY 10158

and by Routledge
4 Park Square, Milton Park, Abingdon, Oxon OX14 4RN

Routledge is an imprint of the Taylor & Francis Group, an informa business

British Library Cataloguing-in-Publication Data
A catalogue record for this book is available from the British Library

Library of Congress Cataloging-in-Publication Data
Names: Begley, Philip, author. | Sheard, Sally, author.
Title: Making genetics and genomics policy in Britain : from personal to
population health / Philip Begley and Sally Sheard.
Description: Abingdon, Oxon ; New York, NY : Routledge, 2023. |
Includes bibliographical references and index.
Identifiers: LCCN 2022013328 (print) | LCCN 2022013329 (ebook) |
ISBN 9781032118383 (hardback) | ISBN 9781032108926 (paperback) |
ISBN 9781003221760 (ebook)
Subjects: LCSH: Genetics--Research--Government policy--Great Britain. |
Genomics--Government policy--Great Britain. | Medical policy--Great Britain. |
Population--Health aspects. | Genomics--Social aspects--Great Britain. |
Genomics--Political aspects--Great Britain.
Classification: LCC QH440 .B44 2023 (print) | LCC QH440 (ebook) |
DDC 576.50941--dc23/eng/20220505
LC record available at https://lccn.loc.gov/2022013328
LC ebook record available at https://lccn.loc.gov/2022013329

ISBN: 978-1-032-11838-3 (hbk)
ISBN: 978-1-032-10892-6 (pbk)
ISBN: 978-1-003-22176-0 (ebk)

DOI: 10.1201/9781003221760

Typeset in Bembo
by MPS Limited, Dehradun

CONTENTS

FIGURES

ACKNOWLEDGEMENTS

The idea for a study of the development of genetics and genomics policy in Britain came from the PHG Foundation. Ahead of their twenty-fifth anniversary in 2022, and at a time when genomic insights continue to be of increasing importance to our collective understanding of medicine and health, they were keen that this important subject should be examined from a historical perspective. We are hugely grateful to them therefore for commissioning us to undertake the research that underpins this book. In addition to financial and practical support, we have received enthusiastic encouragement throughout from Mark Kroese, Philippa Brice, and colleagues. We have also been supported by the members of our Advisory Group – Martin Bobrow, Sir Keith Peters, Diana Walford, and Ron Zimmern – who have provided wisdom and experience and allowed us to draw on their many professional connections. The draft manuscript was read by Roberta Bivins, Frances Flinter, Tony Jewell, and Nick Timmins, who provided valuable feedback.

We would also like to thank all those we have interviewed: Sheila Adam, Mark Bale, Sir John Bell, Martin Bobrow, Stefania Boccia, Angela Brand, Naomi Brecker, Philippa Brice, Wylie Burke, Sir John Burn, Hilary Burton, Sir Mark Caulfield, Tim Cox, George Davey Smith, Dame Sally Davies, Jane Deller, Peter Farndon, Frances Flinter, George Freeman, Alison Hall, Sir Peter Harper, Dame Sue Hill, Layla Jader, Mohamed Karmali, Dianne Kennard, Alastair Kent, Muin Khoury, Bartha Knoppers, Mark Kroese, Anneke Lucassen, Carol Lyon, Eric Meslin, Alan Milburn, Sir Jonathan Montgomery, John Newton, Dame Una O'Brien, Baroness O'Neill, Sharon Peacock, Marcus Pembrey, Sir Keith Peters, Paul Pharoah, Sir Munir Pirmohamed, Sir Bruce Ponder, Ros Skinner, Alison Stewart, Lord Warner, Jacquie Westwood, and Ron Zimmern. Each was engaging and insightful and provided valuable material. Several of them also participated in the two Witness Seminars we held in September and November 2021, the transcripts of which have been published.

Finally, we would like to thank Russell George, Evie Lonsdale and colleagues at Routledge for their support throughout the publication process, friends at the University of Liverpool – particularly Paul Atkinson who has been privy to many conservations about the dynamics of public health genomics over the last couple of years – and our respective families.

Philip Begley and Sally Sheard
Liverpool – May 2022

INTRODUCTION

Revolutions

In 1999, Francis Collins described what a typical clinical encounter might look like in 2010. Collins was the Director of the US National Human Genome Research Institute, and one of the leaders of the international Human Genome Project. The project was set to revolutionise genetic research and thus, in time, the practice of medicine itself.

> John is pleased to learn that genetic testing does not always give bad news—his risks of contracting prostate cancer and Alzheimer's disease are reduced, because he carries low-risk variants of the several genes known in 2010 to contribute to these illnesses. But John is sobered by the evidence of his increased risks of contracting coronary artery disease, colon cancer, and lung cancer. Confronted with the reality of his own genetic data, he arrives at that crucial 'teachable moment' when a lifelong change in health-related behavior, focused on reducing specific risks, is possible.[1]

'John' was a 23-year-old college graduate with high cholesterol. A sample of his DNA had been sent for a series of tests, the results of which gave an insight into his risk of developing a number of common conditions. As an engaged and conscientious patient, John had only agreed to tests for conditions for which suitable preventive interventions were available. These, Collins envisaged, included personally tailored drugs to reduce his cholesterol level and risk of coronary artery disease, annual colonoscopies once John was 45 to check for signs of colon cancer, and support to help him stop smoking in order to address his risk of developing lung cancer. Evoking the words of the influential Canadian physician William Osler, Collins suggested that genetics was the modern means by which medicine

DOI: 10.1201/9781003221760-1

could 'wrest from nature the secrets which have perplexed philosophers in all ages'. 'The genetic revolution', Collins argued, 'is under way'.[2]

This sense of optimism about the impact of genetics on health was shared by others, including in Britain. In 1998, John Bell, Nuffield Professor of Clinical Medicine at the University of Oxford, had looked forward to a point within the next few years when diseases would be understood more in terms of their biomedical mechanisms than their clinical appearance and when genetic variation would be considered alongside other risk factors such as environment.[3] He was confident that genetic testing would be widely used for predicting the development of disease in healthy people, aiding prevention, and that it would facilitate earlier treatments for many patients, including drugs to which they were personally suited. 'Genetic information', Bell argued, 'is likely to transform the practice of clinical medicine'.[4]

While genetics did have an impact on prevention, diagnosis, and treatment over the following decade, the everyday clinical interaction was not transformed in the way that Collins and Bell had expected. Indeed, more than twenty years later, it has not yet been transformed in this way. Bell held on to this vision but later recognised that he had 'got the time frame wrong'.[5] Genetics has, for the moment, powered an evolution in medicine and health rather than a revolution. That this was likely to be the case had always been apparent to observers, both inside and outside the genetics community. The idea of there being transformations in our understanding of disease was not wrong, but such changes would take place incrementally and unevenly over a longer period of time – potentially decades rather than years.[6] Clinicians, bioethicists, and others recognised that different areas of medicine would feel the impact of genetics at different rates. Different healthcare systems would be able to take advantage of genetic insights and innovations in different ways. In a world of limited healthcare resources, there might even be more important priorities. When they came, changes would be contingent on a number of factors – scientific and technological, but also political, economic, and social. There would be significant debates about the potential misuse of genetic information, the danger of discrimination on the basis of genetic predisposition, the need to ensure proper evaluation of new interventions, and the need to have an educated medical workforce.

Collins and Bell had recognised that these issues would be important, though they perhaps underestimated how difficult they would be to resolve. While it was agreed by sceptics that genomic medicine in the round would bring many benefits for clinicians and patients, it would be subject to the vagaries of human behaviour and the pitfalls of translation and implementation in healthcare settings. Clinical interpretation of large amounts of complex genomic data would not be straightforward. Furthermore, genetics would not just be an important subject for clinicians and patients but would influence, and be influenced by, those responsible for the organisation, delivery, and setting of priorities in health services. It also had a distinct history which shaped its subsequent development. By the late 1990s, the stakes may have risen as Collins' revolution threatened to arrive, but many of the issues at the heart of these ongoing changes were not new in themselves.

This book traces the development of public policy in relation to genetics and genomics in Britain. Its focus is the approximate forty-year period – from the 1970s through to the 2010s – during which genetics was transformed from an area of specialist interest to a relatively small group of researchers and practitioners, to a recognised public policy issue with increasing significance across the fields of health and medicine. Genetics initially emerged during the nineteenth and early twentieth centuries as an area of scientific investigation, particularly in terms of understanding the mechanisms of inheritance and the structure of DNA. When it began to have medical implications from the mid-twentieth century onwards, a specialist field of clinical genetics developed. Early leaders were able to shape genetics and its place in research and practice with little central oversight or direction from established health policymakers. Though genetics was emerging as a policy issue, the scope of action of civil servants was limited. Clinical genetics services – as part of Britain's National Health Service, established in 1948 – developed on the ground around the expertise and sphere of influence of key individuals. As new technologies developed and the clinical implications of genetics increased, a number of organisations and professional bodies put forward a more collective position, pushing for the importance of genetics to be more widely recognised and for a greater degree of professional recognition and coordination in the organisation of services. Patient groups also pushed for better provision.

By the 1990s, genetics was increasingly acknowledged as important to general medicine and in relation to common complex conditions as well as established inherited disorders. It began to appear more frequently on the radar of policymakers and politicians. A number of formal public bodies with an interest in genetics began to emerge and shaped the environment in which research and practice took place. By the early 2000s, genetics had arrived as a clear public policy issue in its own right. There was a significant expansion of the policymaking infrastructure around the subject. Funding increased and new biomedical capacity was built. Genetics fitted neatly with many of the wider political and economic preoccupations of the time. The number of areas of medicine in which genetics could facilitate new diagnostics and treatments continued to increase. After a period of consolidation, a further series of important developments followed in the early 2010s as the possibilities of genomic medicine began to be more widely appreciated. Although Collins' revolution is yet to arrive, genetics and genomics now have an established and growing place in the provision of health services and in central policy debates about the future of health and health care

In order to understand these changes, we need to take into account a number of different factors which, often in combination, have been important. First, we need to appreciate the enduring importance of the scientific and technological advances which have driven genetics forward. While early breakthroughs in understanding the chromosomal basis of inherited diseases were important, medical application was relatively limited until sampling techniques were simplified, and diagnostics became easier as part of a gradual shift from the laboratory to the clinic. Important advances in understanding the genetic dimensions of conditions such as thalassaemia and sickle cell anaemia were made during the 1950s and 1960s. From the 1970s, recombinant DNA technology allowed the formation of new sequences, and Sanger sequencing – the first scalable, economic method for reading the whole DNA sequence – facilitated further genetic discoveries and allowed networks of clinical genetics services to be developed across the country. From the 1980s, the successful mapping of particular disease genes, including Huntingdon's disease, Duchenne muscular dystrophy, and cystic fibrosis, the development of the polymerase chain reaction method – which allowed small sections of DNA to be amplified for study, and fluorescence in situ hybridisation – the use of fluorescent probes which to stick to specific places on the human chromosome in order to see if they are present, absent, or duplicated, underpinned shifts towards a molecular understanding of disease.

As a result, clinical genetics was able to broaden out substantially and begin to be integrated into mainstream medicine. It was possible to understand the genetic contribution to an ever-increasing number of inherited disorders, but new technologies also presented the potential of also being able to understand the genetic contribution to common complex conditions. At the same time, the possibilities of genomics started to become apparent. The Human Genome Project was launched in 1990. The race to sequence the whole human genome underpinned further advances such as array CGH testing – through which gains and losses can be ascertained in any part of the genome, and from the late 2000s, next-generation sequencing allowed whole genomes to be sequenced cheaply and quickly. Throughout this period, there has been a steady drumbeat of technological change, punctuated by moments of significant innovation. Nonetheless, the fact that a new technology exists is not in itself transformative. The path from the laboratory to the clinic has not always been straightforward. An extra push has sometimes been required to drive through important developments.

Second, we need to appreciate the political and administrative context in which these changes were taking place. As clinical genetics developed as a speciality, it came to the attention of civil servants, politicians, and other health policymakers. Genetics slowly began to be thought of as a policy issue. However, it was professional organisations such as the Royal College of Physicians and the Clinical Genetics Society that first established what an effective clinical genetics service looked like. For the most part, it was not thought to be the responsibility of policymakers, and there was limited capacity within the Department of Health to take this on. Where there was no real political need for officials to take an interest, for example, in relation to conditions such as thalassaemia and sickle cell anaemia which predominantly affected ethnic minority communities, a number of pioneering clinicians and patient groups pushed for better provision and developed services themselves. Even so, when this picture did begin to change it was the organisation of existing service that was foremost in policymakers' minds. Calls for a new regulatory body to oversee the development of genetics were resisted for a number of years during the 1990s as Ministers were concerned that issues which had not been their responsibility would become more politicised. Yet, public interest in genetics was increasing. Stories about the birth of Dolly the Sheep via cloning, BSE (Bovine spongiform encephalopathy), genetically modified foods, and the forensic uses of DNA, coloured popular perceptions of science.[7]

By the early 2000s, a more detailed understanding of genetics had begun to emerge from scientific and clinical circles and permeate wider debates about social and economic policy. Evidence of more joined-up thinking about the development of services included a recognition that the Department of Health needed to better understand and be prepared for the clinical changes that genetics would bring. A Genetics Policy Unit was formed, and civil servants sought to facilitate collaboration and coordination and make sure that the NHS was able to realise the potential. A more formal policy infrastructure developed as a result. The most high-profile example was the Human Genetics Commission, established in 2000 to advise Ministers and identify priorities in the delivery of services and research. Its members included geneticists, medical practitioners, lawyers, bioethicists, patient representatives, and the pharmaceutical industry. Genetics became a mainstream health policy issue, one in which significant time and money was invested. It was also an issue which could help to frame wider political narratives and resonated with the Labour government's focus on fairness and equity, principles which were thought to be inherent to the NHS. As a result, a number of important policy initiatives were set in motion.

The Conservative/Liberal Democrat coalition government that came to office in 2010 gave new impetus to genetics and genomics policy. The 100,000 Genomes Project rejuvenated policy commitments, with a very personal commitment from the Prime Minister, David Cameron, in 2012 to sequence the genomes of 100,000 NHS patients, an exercise on a scale that had not been attempted before. There has therefore been cumulative and incremental change over time in response to developments in genetics research and practice. A mix of public concern and an appreciation of the opportunities provided a political imperative to act, at important moments, widening the scope of genetics and genomics policy and speeding up its development.

The third factor that influenced genetics and genomics policy was economic imperatives. Elements of competition and cooperation between scientists and researchers at national and international levels have been important throughout the history of genetics, from late nineteenth-century debates about the nature of heredity, to the race to discover the structure of DNA during the 1950s. Moves to ensure that Britain was at the cutting edge of research into the genetics of blood diseases during the 1960s and 1970s reinforced the wider development of molecular medicine. During the late 1980s, British involvement in the Human Genome Project was presented as being scientifically and economically ambitious. In the 1990s, the Conservative government's policy agenda around science and technology identified genetics as a growth area which could boost the British pharmaceutical and healthcare industries and contribute towards wealth creation. During the 2000s, policy debates were often underpinned by an understanding that the development of genetics was bound up closely with British commercial success and economic competitiveness. By the 2010s, rapid developments in genomics were also seen in the context of the government's wider Life Sciences strategy which aimed to boost economic growth and maintain Britain's leading role.[8]

The picture we see in Britain during this period is perhaps best understood in terms of a cumulative or iterative process in which issues around genetics and genomics gained wider recognition before policymakers chose to, or were required, to address them – whether for technological, clinical, political, or economic reasons. Within this, there were several striking examples when it was the intervention of key individuals that moved issues to the top of the policy agenda. When this has occurred, it has often been disruptive and controversial. Such moments of impetus are not without cost, but they have been important in moving the conversation forward, changing the terms of the debate, or speeding up the process of translation and implementation.

In the course of examining these developments, this book takes a particular interest in the role of 'public health genomics'. While it has been described as a field, a discipline, or a body of knowledge, those in and around public health genomics have most often conceptualised it as an enterprise or a collective way of approaching problems. The collective definition agreed in 2005 was that public health genomics concerned 'the responsible and effective translation of genome-based knowledge and technologies for the benefit of population health'.[9] Public health genomics has its origins in historic intersections between the interests of geneticists and public health experts. These manifested themselves in newborn screening programmes for conditions such as Phenylketonuria from the 1960s, and then more formally in the field of genetic epidemiology, which sought to incorporate new genetic and statistical approaches into the longstanding traditions of epidemiology.

The label 'public health genetics' first emerged in the United States during the 1990s when a small group of public health physicians, geneticists, and bioethicists sought to foster more philosophical debates about the potential implications of genetic and genomic medicine. They sought to confront difficult ethical and legal questions as well as practical ones while seeking to ensure the widest possible population health benefits. Key individuals such as Muin Khoury in the United States and Ron Zimmern in Britain were able to build on these foundations. Khoury, from 1997, led the Office for Genomics and Public Health within the US Centers for Disease Control and Prevention (CDC), and Zimmern, a public health physician, established the Public Health Genetics Unit (PHGU) in Cambridge in the same year. His aim was to stimulate a different kind of conversation amongst practitioners and policymakers, and to enter debates about research, the development of genetic services and genetics policy from the perspective of a non-geneticist. Many of the individuals that shaped public health genetics were initially trained in and practiced public health medicine. These experiences often instilled a set of values and a perspective that stayed with them through their careers, founded on an appreciation of the value of population-wide approaches to health and healthcare-related questions.

International collaboration was also central to public health genomics. At a key meeting in Bellagio, Italy, in 2005, it was decided by a core group in and around the field to use this phrase rather than the potentially narrower conception of 'public health genetics'. The approach was consciously broad so as to encompass all disease-causing gene interactions. Public health genomics was thought to have the potential to stimulate a balanced public debate. It could consider disease prevention at different levels – before a disease developed and once it had developed – and offer a bridge between individual health and population health, ensuring a solid evidence base for the use of genetic tests, screening programmes, and other interventions. At the same time, it could help to develop appropriate regulatory and public policy frameworks which considered economic, legal, ethical, and social factors as well as scientific knowledge.

As we explore in this book, many of the ideas and approaches inherent to public health genomics have had an important place in debates about the development of genetics and genomics policy in Britain. Their influence has often been diffuse and difficult to pin down, but it is appreciable. The Public Health Genetics Unit, and its successor the PHG Foundation, have formed a critical part of the emerging, supportive policy infrastructure around genetics, building influential connections, and disseminating ideas. Many of their reports and research projects are acknowledged by the policy, academic, and service communities as significant achievements. However, the main contribution of public health genomics is evident in the increasing numbers of practitioners and policymakers who appreciate the importance of genome-based factors alongside environmental factors in understanding human disease, and who appreciate the value of a wider view that includes economic, legal, and social factors, particularly in the British context of implementing genetic and genomic medicine in the NHS.

Approaches

In contrast to most of the individuals encountered in this book, the authors are historians by training. The research on which our book is built has therefore been carried out using established historical methodologies.[10] Although histories of genetic science and the development of clinical practice have been written, particularly by the late Sir Peter Harper, to date, there has not yet been a detailed study of genetics from a British policy perspective.[11] Similarly, the history of public health genomics has been considered by those in and around the field, but not in analytical detail.[12] We hope that this book fills the gap and informs ongoing policy debates. In constructing this history, we have drawn on a rich set of written sources, including policy documents and reports produced by professional organisations, regulatory bodies, and government departments, newspapers, periodicals and journal articles, and the records of Parliamentary debates. The PHG Foundation has also kindly allowed us access to internal correspondence, documents from research projects and courses, and information relating to conferences, seminars, and meetings.

At the heart of this book, however, is 'oral history'.[13] We conducted 47 research interviews with a range of individuals from across the organisations, institutions, and disciplines discussed in the chapters which follow. These included clinical geneticists, bioethicists, public health practitioners, physicians, politicians, civil servants, and other academics and researchers. The individuals come from Britain, the United States, Canada, and Europe. We have also drawn on interviews conducted by Sir Peter Harper as part of the Genetics and Medicine Historical Network and from the Cold Spring Harbor Laboratory Oral History Collection, which are available online.[14] Oral history is particularly valuable as a research method because it can bring out details and perspectives difficult to appreciate through written sources alone. We also held two 'witness seminars' in September and November 2021. Witness seminars allow memories to be shared and contested in order to construct a useful analysis of past events. We first brought together a group of individuals who were directly involved in the development of genetics and genomics policies in Britain or who had experience of their impact. The witnesses discussed issues such as when and why health policymakers developed an interest in genetics, when the focus began to change from genetics to genomics, and the key milestones in the emergence of genomic medicine. The second witness seminar focussed more explicitly on public health genomics – its history, the nature of international collaborations, and its influence. The transcripts of both events have been published together.[15]

We have aimed to make this study detailed but also accessible to a wide audience. We hope that it will be read not only by those involved in the research and practice of genetics, genomics, bioethics, and population health, but also by NHS staff, policymakers, politicians, and the public. Only by understanding how we got to where we are today, can we fully begin to appreciate what our future policy options might be, and their implications. Whilst many of the concepts discussed in the course of this book are well known and relatively straightforward – many readers will recall Mendel's peas and the basic principles of inheritance from school science lessons – some are more complex. After careful consideration, we decided not to include too much detail about the underlying science of genetics and genomics. This is available in other sources – for example, *Genetics, Health Care and Public Policy: An Introduction to Public Health Genetics*, published by representatives of the PHG Foundation and guides such as *Genes: A Very Short Introduction*.[16] There also are many introductory textbooks on genetics which can provide further detail.[17] The aim here has been to include enough information in the text – about new genomic techniques, for example – to allow a lay reader to follow the narrative, without going into intricate detail about exactly how they work. Here we provide here a brief glossary of key terms which are discussed frequently throughout the book.

Definitions

Genetics: The study of genes and their role in inherited characteristics.

Clinical Genetics: The study of inherited genetic disorders, their diagnosis and treatment.

Cytogenetics: The study of chromosomes and DNA and their role in inheritance.

Genetic Determination: The idea that human behaviour and characteristics are directly determined by an individual's genes, at the expense of the role of the environment and other factors.

Genetic Discrimination: The act of treating someone less favourably – often by an employer or an insurer – because they have a genetic predisposition to a particular disease or condition.

Genomics: The study of the whole genome – all of an individuals' genes, all the DNA inside each cell – and the interactions between those genes.

Whole-Genome Sequencing: The technological process of constructing an individual or organism's whole genome so that it can be analysed.

Public Health: A discipline and a practice – the study of, and efforts to improve, the health of the population or populations, as a whole – incorporating elements such as epidemiology, disease prevention, and health protection.

Population Health: The study of health outcomes, their patterns, and their determinants amongst a group of individuals, often at a national level.

Epidemiology: The study of the causes and patterns of disease in populations.

Public Health Genomics: Originally *Public Health Genetics* until 2005. A field of study or collective approach which aims to facilitate the effective translation and implementation of genetic and genomic knowledge in order to improve population health.

Policy: The position on a given issue held by an organisation, institution, or government department.

Policymakers: Those responsible for the development of relevant policies – for example, civil servants, politicians, and professional representatives.

Structure

Chapter 1 – *Origins* – provides context for the chapters that follow. It establishes the relevant history of genetics as an area of scientific investigation, particularly in terms of understanding the mechanisms of inheritance and the structure of DNA. It then considers the ways in which genetics came to have implications for medicine, leading to the development of the specialist field of clinical genetics. The chapter places these developments alongside an account of the development of public health as a discipline and practice and establishes how shared interests and approaches between genetics and public health provided a foundation on which subsequent fields such as public health genomics were able to build. We then examine the emergence of genetics as a policy issue, tracing official interest from the 1970s. The chapter then concludes by describing the first phase of development of public health genomics in Britain in the 1990s, highlighting the contribution of Ron Zimmern, Director of Public Health for Cambridge and Huntingdon Health Authority.

Chapter 2 – *Foundations* – establishes the importance of ELSI issues – economic, legal, and social issues within this history, particularly through the Human Genome Project which began in 1990. It draws attention to the work of a number of influential international organisations, including the Human Genome Organization and UNESCO. We then consider the extent to which clinical geneticists in Britain drew on an increasingly supportive policy infrastructure from the mid-1990s, and a series of official and unofficial reports which made the case for greater coordination of genetic services and highlighted the potential impact of genetics on mainstream medicine. The chapter also describes the continuing development of public health genomics and its first steps towards formalisation as a field, including the establishment in 1997 of the Public Health Genetics Unit in Britain and the CDC Office of Public Health in the United States, and the building of international collaborations.

Chapter 3 – *Progress* – focuses on the steps taken by the Public Health Genetics Unit to build interest in its work, including making connections with policymakers and publishing influential reports. By the early 2000s, important elements of their approach were beginning to feed through into policymaking discussions. This chapter charts how genetics moved up the central policy agenda through the creation of new public bodies, and a Genetics White Paper in 2003 helped to set a number of key initiatives in motion. In 2003, the completion of the Human Genome Project then demonstrated the real and immediate potential of genomics.

Chapter 4 – *New Directions* – discusses the importance of international collaborations in public health genomics and the agreement of a shared analytical approach. In Britain, there was a period of consolidation without further major policy announcements in parallel with further scientific developments, including the speeding up and industrialising of whole-genome sequencing and the first stages of meaningful provision in genomic medicine. The 2009 House of Lords Science and Technology Committee report *Genomic Medicine* led to the government establishing a high-profile Human Genomics Strategy Group. However, outside of the usual policymaking processes, impetus was given to the 100,000 Genomes Project in 2012. We set out how this was conceptualised and its role in consolidating genomics as a core part of the government's integrated life science and health service strategies.

The *Conclusion* then summarises the book and further details each of the main points around the development of genetics as a public policy issue and the place of public health genomics. We reflect on the implications of this history and its significance for informing future developments and debates, for example, around the likely place of genomics in public health practice and personalised health care. While the focus of this book has been on policy development and process at a high level, the conclusion also highlights important issues that form part of this story that future researchers will be able to consider.

Notes

1 F.S. Collins, 'Medical and Societal Consequences of the Human Genome Project', *New England Journal of Medicine*, Vol. 341, No. 1, 1999, p. 35. Based on his Shattuck Lecture to the Annual Meeting of the Massachusetts Medical Society.

2 Ibid, p. 36.

3 J. Bell, 'The New Genetics in Clinical Practice', *British Medical Journal*, Vol. 316, February 21, 1998, pp. 618–620.

4 Ibid, p. 618.

5 G. Watts, 'Professor Sir John Bell, President of the Academy of Medical Science', *Clinical Medicine*, Vol. 9, No. 5, 2009, p. 463.

6 J.P. Evans, E.M. Meslin, T.M. Marteau, and T. Caulfield, 'Deflating the Genomics Bubble', *Science*, Vol. 331, No. 6019, 2011, p. 861–862.

7 Dozens of patients died in Britain during the 1980s and 1990s after eating infected beef and developing Creutzfeldt–Jakob disease, the human equivalent of Bovine spongiform encephalopathy, known colloquially as 'mad cow disease'. See, for example, G. Rivett, *From Cradle to Grave: Fifty Years of the NHS* (King's Fund, 1998).

8 *Strategy for UK Life Sciences* (Department for Business, Innovation and Skills, 2011).

9 *Genome-Based Research and Population Health*, Report of an expert workshop held at the Rockefeller Foundation Study and Conference Centre, Bellagio, Italy, 14–20 April 2005, p. 7.

10 J. Tosh, *The Pursuit of History: Aims, Methods and New Directions in the Study of History* (Routledge, 2022). L. Jordanova, *History in Practice* (Bloomsbury, 2019). J.L Gaddis, *The Landscape of History: How Historians Map the Past* (Oxford University Press, 2002).

11 P.S. Harper, *The Evolution of Medical Genetics: A British Perspective* (CRC Press, 2020). H.I. Petermann, P.S. Harper, and S. Doetz (eds.), *History of Human Genetics: Aspects of its Development and Global Perspectives* (Springer, 2017).

12 C.M. Molster, F.L. Bowman, G.A. Bilkey, A.S. Cho, B.L. Burns, K.J. Nowak, and H.J.S. Dawkins, 'The Evolution of Public Health Genomics: Exploring its Past, Present and Future', *Frontiers in Public Health*, Vol. 6, No. 247, 2018. M.J. Khoury, M. Scott Bowen, M. Clyne, W.D. Dotson, M.L. Gwinn, R.F. Green, K.Kolor, J.L. Rodriguez, A. Wulf, and W. Yu, 'From Public Health Genomics to Precision Public Health: A 20-Year Journey', *Genetics in Medicine*, Vol. 20, No. 6, 2018, pp. 574–582.

13 L. Abrams, *Oral History Theory* (Routledge, 2016). K. Howarth, *Oral History* (Sutton, 1999). V. Berridge, 'Hidden from History?: Oral History and the History of Health Policy', *Oral History*, Vol. 38, No. 1, 2010.

14 https://genmedhist.eshg.org/. https://library.cshl.edu/oralhistory/

15 *The Development and Influence of Public Health Genomics* (University of Liverpool, 2022).

16 A. Stewart, P. Brice, H. Burton, P. Pharoah, S. Sanderson, and R. Zimmern, *Genetics, Health Care and Public Policy: An Introduction to Public Health Genetics* (Cambridge University Press, 2007). J.M.W. Slack, *Genes: A Very Short Introduction* (Oxford University Press, 2014).

17 W.S. Klug et al., *Concepts of Genetics* (Pearson, 2012). P.D. Turnpenny, S. Ellard, and R. Cleaver, *Emery's Elements of Medical Genetics* (Elsevier, 2021). T. Brown, *Introduction to Genetics: A Molecular Approach* (Garland, 2012).

Bibliography

Abrams, L., *Oral History Theory* (Routledge, 2016).

Bell, J., 'The New Genetics in Clinical Practice', *British Medical Journal*, Vol. 316, 21 February 1998.

Berridge, V., 'Hidden from History?: Oral History and the History of Health Policy', *Oral History*, Vol. 38, No. 1, 2010.

Bivins, R., 'Coming 'Home' to (Post) Colonial Medicine: Treating Tropical Bodies in Post-War Britain, *Social History of Medicine*, Vol. 26, No. 1, 2013.

Bivins, R., *Contagious Communities: Medicine, Migration, and the NHS in Post War Britain* (Oxford University Press, 2015).

Brown, T., *Introduction to Genetics: A Molecular Approach* (Garland, 2012).

Collins, F.S., 'Medical and Societal Consequences of the Human Genome Project', *New England Journal of Medicine*, Vol. 341, No. 1, 1999.

Evans, J.P., Meslin, E.M., Marteau, T.M., and Caulfield, T., 'Deflating the Genomics Bubble', *Science*, Vol. 331, No. 6019, 2011.

Gaddis, J.L., *The Landscape of History: How Historians Map the Past* (Oxford University Press, 2002).

Genome-Based Research and Population Health, Report of an expert workshop held at the Rockefeller Foundation Study and Conference Centre, Bellagio, Italy, 14–20 April 2005.

Harper, P.S., *The Evolution of Medical Genetics: A British Perspective* (CRC Press, 2020).

Howarth, K., *Oral History* (Sutton, 1999).

Jordanova, L., *History in Practice* (Bloomsbury, 2019).

Khoury, M.J., Scott Bowen, M., Clyne, M., Dotson, W.D., Gwinn, M.L., Green, R.F., Kolor, K., Rodriguez, J.L., Wulf, A., and Yu, W., 'From Public Health Genomics to Precision Public Health: A 20-Year Journey', *Genetics in Medicine*, Vol. 20, No. 6, 2018.

Klug, W.S., et al., *Concepts of Genetics* (Pearson, 2012).

Molster, C.M., Bowman, F.L., Bilkey, G.A., Cho, A.S., Burns, B.L., Nowak, K.J., and Dawkins, H.J.S., 'The Evolution of Public Health Genomics: Exploring its Past, Present and Future', *Frontiers in Public Health*, Vol. 6, No. 247, 2018.

Petermann, H.I., Harper, P.S. and Doetz, S. (eds.), *History of Human Genetics: Aspects of its Development and Global Perspectives* (Springer, 2017).

Rivett, G., *From Cradle to Grave: Fifty Years of the NHS* (King's Fund, 1998).

Slack, J.M.W., *Genes: A Very Short Introduction* (Oxford University Press, 2014).

Stewart, A., Brice, P., Burton, H., Pharoah, P., Sanderson, S., and Zimmern, R., *Genetics, Health Care and Public Policy: An Introduction to Public Health Genetics* (Cambridge University Press, 2007).

Strategy for UK Life Sciences (Department for Business, Innovation and Skills, 2011).

Tosh, J., *The Pursuit of History: Aims, Methods and New Directions in the Study of History* (Routledge, 2022).

Turnpenny, P.D., Ellard, S., and Cleaver, R., *Emery's Elements of Medical Genetics* (Elsevier, 2021).

Watts, G., 'Professor Sir John Bell, President of the Academy of Medical Science', *Clinical Medicine*, Vol. 9, No. 5, 2009.

1

ORIGINS

The History of Genetics

Genetics emerged as a scientific and medical field and an area of health policy over the course of the twentieth century, stimulated by advances in scientific knowledge, changes in the provision of healthcare and conceptual shifts in the discipline and practice of public health. The first observations of the genetic differences between humans have been traced to 5000-year-old Babylonian clay tablets.[1] Many of the basic principles were apparent to physicians and scholars in Ancient Greece.[2] Some theorised that the hereditary information passed between generations was carried within the sperm, although Aristotle pointed out that this couldn't explain how children inherited characteristics from their mothers as well as their fathers. Even so, this first approach underpinned understandings of reproduction for many centuries, including the popular idea of the 'homunculus' – a tiny human that moved fully formed into a mother's womb.[3]

During the early-modern period, scientists and physicians with an interest in biological inheritance began to lay the foundations for modern genetics. In the mid-seventeenth century William Harvey – best known for his description of the circulation of the blood – observed embryonic development in animals, and by the 1700s John Dalton had begun to record patterns of colour blindness amongst his family.[4] Though unknown at the time, Dalton's work would later be recognised as demonstrating a classic case of X-linked inheritance when a trait is passed on from one generation to the next through a mutation on the X chromosome. Studies of the differences between disposition and predisposition to disease – whether a disease could simply be explained by inheritance or if an environmental factor was also needed – by Joseph Adams in the early 1800s, and the descriptions of muscular dystrophy recorded by Edward Meryon and Guillaume-Benjamin-Amand Duchenne during the 1860s, are early examples of thinking about heredity in the

DOI: 10.1201/9781003221760-2

realm of medicine.[5] Adams has been described as 'the first clinical geneticist' and a 'forgotten founder of medical genetics'.[6]

The work of Charles Darwin and contemporaries such as Alfred Russel Wallace from the 1850s on evolution and natural selection was also important. Established theories of heredity at the time were underpinned by relatively simple observations of 'blended' inheritance – the fact that offspring appeared to share a range of traits from both of their parents. Darwin delivered the first 'credible mechanism' for understanding such observed changes, by explaining inheritance through the theory of 'Pangenesis' – 'the beginning of everything' – the idea that each part of the body emits small particles of itself (which he named 'gemmules'), that move to the reproductive organs and are then transferred to the offspring during reproduction.[7] The central hypothesis was revolutionary, but it was difficult to reconcile with observations of blended inheritance. Critics such as Fleeming Jenkin questioned why traits did not disappear over time if they were seemingly diluted with each new generation. Darwin's theory also failed to account for the phenomenon of variation – where offspring exhibited characteristics not visible in either parent. Darwin acknowledged that he could not explain this. Nonetheless, when *On the Origins of Species by Means of Natural Selection* was published in November 1859 (and all 1250 copies sold on the first day), it was clear that a fundamental shift in ways of thinking about inheritance was underway.

It was the Austrian monk (and highly skilled scientist) Gregor Mendel, who provided the key breakthrough while studying natural sciences at the University of Vienna. Mendel started breeding peas as a hobby in the 1850s. Breeding of plants and animals was a common agricultural practice, drawing on centuries of knowledge about how characteristics could be bred in and bred out through controlled fertilisation. Mendel collected 34 strains of peas from local farmers and bred them to get 'true' offspring – identical in terms of seed shape, flower colour, and height of the plant. He developed what would now be thought of as an experimental protocol, carefully recording what happened when he manually cross-pollinated specific colours and sizes.

Mendel recognised that this was 'fixed' information being passed between generations, with equal amounts of 'information' provided by the male and the female parent plants. He also recognised that there were 'dominant' and 'recessive' traits, which explained why some characteristics appeared to skip generations.[8] After eight growing seasons, Mendel had amassed a significant set of data from 28,000 pea plants. He presented his findings to the Brno Natural Science Society in 1865, describing the mechanism of heredity for the first time in a paper in the Society's journal the following year. Copies were sent to English-speaking scientific societies. However, it received only four citations before being 'rediscovered' in 1900, and Darwin, who was wrestling with his own theory at the time, appears to have missed a reference to it in a book on plant hybrids that he read in the 1870s. Mendel had tried in vain to collaborate with other German-speaking researchers but found his approaches rebuffed because he came from outside established academic circles. He died in 1884.

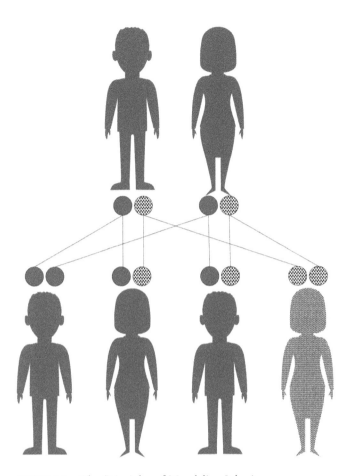

FIGURE 1.1   The Principles of Mendelian Inheritance

Source: Reproduced with the permission of the PHG Foundation.

With Mendel's work going largely unrecognised, scientists continued to look for answers. During the 1890s, the German embryologist August Weissmann developed the 'germplasm' theory, identifying that, in contrast to somatic cells throughout the body, only germ cells in the reproductive organs can transmit heritable information. The Dutch botanist Hugo de Vries, stimulated by discussions with Darwin before his death in 1882, and also without knowledge of Mendel's work, developed the concept of 'pangenes' – the idea that each characteristic or trait was transmitted by tiny particles. When a friend did send de Vries an old copy of Mendel's 1865 paper he realised that he needed to get his own research published as quickly as possible, and initially failed to acknowledge what he had gained from reading Mendel's work. De Vries' paper was one of three in 1900 that drew on Mendel's research – the other scientists had similarly stumbled across the 1865 paper on pea hybrids. De Vries later admitted: 'Modesty is a virtue, yet one gets further without it'. As Siddhartha

Mukherjee notes, 'Being rediscovered once is proof of a scientist's prescience. Being rediscovered thrice is an insult'.[9]

Serendipity also appears to have played a role in the dissemination of Mendel's work in Britain. The Cambridge biologist William Bateson is reported to have read de Vries' paper in May 1900 on the train to London where he was due to give a talk on heredity to the Royal Horticultural Society.[10] Once he knew of Mendel's work, Bateson re-shaped his own research and went to such lengths to correct the scientific injustice that he gained the nickname 'Mendel's bulldog'. We can see therefore that far from being unique to the late twentieth and early twenty-first centuries, cultures of national and international competition and cooperation between scientists and practitioners have been important throughout the history of genetics. Bateson realised that he needed a term to describe the new scientific field that was developing. His initial attempt – 'pangenetics' was unhelpfully loaded with Darwinian associations. In 1906 therefore, he simplified it to 'genetics': from the Greek 'genno' – 'to give birth'. Bateson recognised the Pandora's box that was being opened: 'What will happen when … the facts of heredity are … commonly known? … One thing is certain: mankind will begin to interfere'.[11] He was one of a growing number of scientists who drew with increasing confidence on evidence of hereditary. But there was conflict between them.

From the late nineteenth century, Darwin's radical theory of the 'survival of the fittest' had been exploited by a group of gentlemen-scholars, led by his cousin Francis Galton. Galton's own scientific studies and publications were thin in comparison with Darwin's, but they gave him some credibility with scholars interested in measuring human characteristics such as height and eye colour, and amongst those who were also attempting to measure intelligence through less objective proxies such as educational achievement. Galton coined the term eugenics in 1883 – a blend of the Greek for good: 'eu', and genesis: 'origins' – to speak to the importance of *nature* over *nurture,* drawing on what he claimed to have observed in genealogical studies of the landed gentry. His 'Law of Ancestral Heredity' suggested that around 50% of a child's inherited traits came from their parents, 25% from their grand-parents, and so on. Key associates of Galton included the mathematician Karl Pearson, who, around the turn of the twentieth century – a period marked by conceptions of white racial superiority and British colonialism – often talked about eugenics in terms of historical racial and national struggles.[12] Pearson built on the studies of the Belgian scientist Adolphe Quetelet which demonstrated the 'normal' bell-curve distribution of features such as height and weight within populations.

During the early twentieth century, at venues such as the Royal Society of Medicine in London, Galton and his eugenicist supporters entered into fierce debates with scientists led by Bateson, Ronald Fisher, and others who held the line that inherited information was carried as fixed units from parents, not 'by halved or quartered messages from ghostly ancestors'.[13] It was the Danish botanist Wilhelm Johannsen who gave these units the name 'genes' in 1909. According to Johannsen, 'It expresses the only evident fact that … many characteristics of the organism [are] specified … in unique, separate and thereby independent ways'.[14] Resolution was eventually achieved as the scientific community agreed that the philosophies of evolution and natural selection were compatible with Mendelian inheritance via multiple fixed units of heredity. In 1942, Julian Huxley described this fusion, and the field of 'evolutionary biology' which had developed as a result, as the 'modern synthesis'.[15]

Bateson was also a key figure in recognising the implications of Mendelian inheritance for human disease. Genetic disorders that previously could only be observed were, thanks to this intense period of research and debate, now clinically understandable, and therefore potentially treatable. In 1900 the Austrian biologist Karl Landsteiner discovered the ABO blood group system – a variable human characteristic that followed Mendelian rules. In 1902, the British physician Archibald Garrod (a close associate of Bateson) made a key breakthrough in the understanding of Alkaptonuria – a human inherited disease that followed Mendelian rules. This approach was taken forward by figures such as Fisher, J.B.S Haldane, and Lionel Penrose. Haldane in particular is credited with further exploiting the mathematical dynamics of genetics and undertaking many of the first genetic linkage studies – through which it could be shown that genetic traits in close proximity on a chromosome could be inherited together – including demonstrating the effect with regards to haemophilia in 1937.[16]

Such advances in scientific understanding, however, also stoked ongoing eugenic debates about the wider social implications. During the early decades of the twentieth century, there was increasing concern about the falling birth-rate in Britain and the health and fitness of the population, in both moral and physical terms.[17] Galton founded the Eugenics Education Society in 1907 (later the Galton Institute) and in 1909, a journal, *The Eugenics Review*. The journal publicised proposals not just for selective breeding but also for sterilisation of the 'feeble-minded'. Yet many leading scientists and geneticists, including Penrose, Haldane, and the biologist Lancelot Hogben, opposed such ideas during the inter-war period, preventing a eugenic culture from strongly influencing the direction of genetics research in Britain.[18] The Second World War subsequently severely disrupted many of the patterns of research and innovation that had supported the growth of human genetics in Britain. When the war ended, however, there were still strong foundations in place for future development.

As well as shaping understandings of the patterns and effects of heredity, the nineteenth century was a formative period in the study of its mechanisms and DNA. The Swiss biologist Friedrich Miescher first identified nucleic acids in cell nuclei in 1869 and laid the groundwork for future understanding of their role in inheritance.[19] Important advances were then made by the German biologist Walther Flemming, who identified a threadlike structure inside each cell nuclei (later named chromosomes) and established the process of cell division (mitosis), identifying that all cell nuclei are derived from that of a parent cell.[20] The fact that there is a characteristic chromosome number for each species, and that it halves during the process of meiosis, was first identified by the Belgian embryologist

Edouard van Beneden.[21] The German biochemist Albrecht Kossel subsequently analysed the chemical composition of nucleic acid and identified the five bases – adenine, thymine, cytosine, guanosine, and uracil – of the molecular structure of DNA and RNA.[22] The American biochemist Phoebus Levene established that sugar ribose and deoxyribose were components of nucleic acid and that molecules consisted of bases, sugars, and phosphates joined together in a long chain.[23] Further important steps were taken by the American biologist Thomas Hunt Morgan who, through research with Drosophila fruit flies, mapped genes, successfully identified that they were carried on the chromosome, and established their role in heredity.[24]

The 1944 study by the Canadian-American microbiologist Oswald Avery, and his colleagues Colin Macleod and Maclyn McCarty, which identified that it was DNA that carried genetic information, rather than cell protein as had been

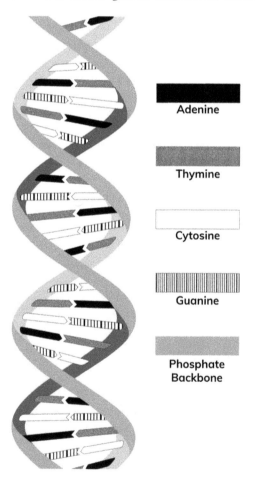

FIGURE 1.2 The Double Helix Structure of DNA

Source: Reproduced with the permission of the PHG Foundation.

assumed, was a hugely important breakthrough. In 1950 the Austro-Hungarian born American biochemist Erwin Chargaff outlined two key rules, which set out that the bases (guanine and cytosine and adenine and thymine) occur in a 1:1 ratio and are paired. This understanding ultimately paved the way for the discovery of the base pair double helix structure of DNA.

The race to identify the structure of DNA – a further example of the importance of international competition and collaboration – was headlined by a team in the United States (led by Linus Pauling), a team at King's College, London (led by Maurice Wilkins and Rosalind Franklin), and a team at the University of Cambridge (led by James Watson and Francis Crick). The Cambridge team focussed on building molecular models, while the KCL team worked with X-ray diffraction – a technique that allowed the atomic structure of materials to be observed. The most important image they produced was the famous Photo 51, taken by Rosalind Franklin and Raymond Gosling. After Wilkins subsequently shared elements of the KCL group's work, Crick and Watson were able to use their insights to complement their own research and conceptualise the DNA double helix structure. The key papers were published in *Nature* in April 1953.[25] Crick, Watson, and Wilkins shared a Nobel Prize in 1962, while Franklin and others went relatively unrecognised for several decades.[26] Once the breakthrough was made however, the gene became a central, unifying concept across the life sciences.

The Development of Medical Genetics

The work of early geneticists in Britain provided a strong foundation for the field to rebound after the Second World War. Many had medical interests and much of their work would have significant medical implications. Between 1932 and 1939, the Medical Research Council – the central body responsible for distributing medical research funding in Britain – had also started to take an interest in genetics, through the formation of a Human Genetics Committee.[27] Yet, the links between genetics and medicine remained underdeveloped. The emergence of medical genetics as a distinct area of activity, and ultimately a speciality in its own right was a gradual process, driven largely by pioneering individuals in the field.[28] Sir Peter Harper, the leading historian of medical genetics in Britain, and also one of its most important practitioners for many years, traced its emergence back to two key developments – the research done at University College London, and particularly its Galton Laboratory led by Lionel Penrose, and that conducted by a group of research units established by the Medical Research Council in response to Cold War concerns about the genetic effects of atomic radiation.[29]

Harper saw Penrose as 'the person who, more than anyone else was responsible for determining the shape of British human genetics and, though less directly, what would over the coming decades become medical genetics'.[30] This was due in large part to Penrose's determined efforts to disconnect genetics, particularly at UCL, from any association with eugenics, and his bringing together of a group of researchers, including regular international visitors, who themselves became leaders in the field. He was also known to have a caring interest in patients and their families, particularly those with Down's syndrome.[31]

Though the development of a new field of medical genetics was not the express intention, the MRC also played an important role by sponsoring a number of units for genetic research in Edinburgh, Harwell, Oxford, and London during the late 1940s and 1950s.[32] Influential leaders included Alan Stevenson at the Clinical and Population Genetics Unit in Oxford, and Paul Polani at Guy's Hospital, London. It was in Edinburgh, at what was originally known as the 'Clinical Effects of Radiation Unit', led by Michael Court Brown, that Patricia Jacobs and colleagues first identified the link between Trisomy 21 – having an extra copy of the twenty-first chromosome – and Down's syndrome, and other sex chromosome abnormalities. The Harwell unit was instrumental in improving understanding of the nature of Turner syndrome – where one or part of one of the X chromosomes is missing in a female. According to Harper, 'This dramatic series of chromosomal findings, all published in 1959 and 1960, and featuring principally UK-based workers, firmly established human cytogenetics [the study of the structure of DNA] at the forefront of human genetics generally'.[33] Karyotyping (photographing or testing to determine an individual's chromosome complement), though an important step forward and a technique that had been used in many breakthroughs, remained the main technology that was available for many years, and the early focus was often on research into the causes of inherited diseases rather than looking for medical applications. The leading research centres were not yet well connected with universities and the wider NHS in a way that might have allowed new services to develop.

The first physician to attempt to place genetics close to the heart of his practice was Cyril Clarke in Liverpool during the 1950s and 1960s. Clarke is still best known for his work on rhesus haemolytic disease – a condition that occurs when there is an incompatibility between the blood types of a mother and a foetus, and maternal antibodies cross the placenta and attack the red blood cells of the foetus. Alongside his colleague Ronald Finn, Clarke provided a workable intervention via maternal immunisation of immunoglobin during pregnancy.[34] The reasons why one or both were never awarded a Nobel Prize is still the subject of debate amongst the medical research community. Many of the leading medical geneticists in Britain worked with Clarke early in their careers and were influenced by his approach, which included trying to ensure that clinicians had a solid medical training before beginning to think about genetics. Members of this 'Liverpool diaspora' included Peter Harper, David Price Evans, Rodney Harris, Marcus Pembrey, and David Weatherall. Weatherall – later Nuffield Professor of Clinical Medicine and Regius Professor of Medicine at the University of Oxford – once recalled, 'When I was Cyril Clarke's houseman in 1956, medical genetics was just starting to zoom, lots of excitement, everybody rushing around the hospital spitting into tubes and smelling their urine, and lots of very bright young people coming along for research projects'.[35] He and others in Liverpool also gained valuable experience through working with Victor McKusick at Johns Hopkins University Hospital in Baltimore in the United States. McKusick was a close friend of Clarke and has been widely recognised as a 'founding father' of medical genetics.[36]

His influence was felt once those who had studied in the United States came back in Britain. Harper recalled that 'in places like Manchester and Glasgow with Malcolm Ferguson-Smith, and to some extent in Edinburgh with Alan Emery, myself in Cardiff, London with Marcus Pembrey, we set up things, which actually in my case was more or less a sort of replica of what was happening at Baltimore'.[37]

Yet Martin Bobrow, then a young doctor, saw the field in the 1960s as still 'pulling itself up by its shoestrings' and something of a 'wild west scene'.[38] There was no established training and no clear career path. Bobrow had arrived in Britain having studied medicine in South Africa: 'There wasn't a way into genetics at that stage … there were probably three places in the UK that genetics was being practiced within a medical context … so if you didn't get a job with one of those three there wasn't anywhere else to go'.[39] Bobrow was fortunate to find a position at the MRC genetics unit in Oxford and went on to be a leader in the field. The same was true for Peter Farndon who, following advice from Cedric Carter – 'get yourself grounded as a proper clinician, because nobody believes that geneticists are proper physicians' – gained experience in general medicine before specialising in genetics, and taking up one of the few senior registrar posts in the country in Manchester, before joining the University of Birmingham as Professor of Clinical Genetics in 1984.[40] Alan Johnston, who went on to be Clinical Senior Lecturer in Medicine and Genetics in Aberdeen, has also noted the limited opportunities: 'In 1960 I had just come back from spending over a year with Victor McKusick. I was full of enthusiasm, but there was no funding for clinical work by the authorities for me to become a clinical geneticist … I was told to go back to work as a senior medical registrar … There was definitely a financial element: laboratory work could be funded but not clinical work'.[41] For much of the 1950s and 1960s the work of medical genetics was largely theoretical. The science of DNA was still developing and there was little opportunity for practical application.

From the late 1960s, however, scientific, and technological advances, as well as organisational efforts, helped to move medical genetics forward. Chromosomal analysis and diagnostics became easier. Sampling techniques became less invasive, allowing more samples to be taken and demand for new specialist services began to rise. An important example was molecular prenatal diagnosis during the first trimester of pregnancy, which was developed in relation to thalassaemia and other haemoglobinopathies by Bernadette Modell, a Senior Paediatric Registrar at University College Hospital in London, alongside carrier screening and other vital services for minority ethnic communities which helped to reduce the risk of miscarriage and the need for late terminations.[42] There was a gradual shift in focus from the laboratory to the clinic as a result. According to Bobrow: 'This was the territory that genetics then stamped out for itself: interpreting laboratory results, some of which were rather complex, and even pretty good paediatricians – they were all paediatric at that time – found it relatively hard to get to grips with the technology quickly'.[43] The US historian Susan Lindee identifies this as a crucial 20-year period: 'there were enough people and institutions around with enough stakes in a technological conception of genetic disease to produce a series of events that have had lasting and profound consequences for the development of bio-medicine'. Lindee sees this as a defining period during which 'human genetics was transformed from a medical backwater to an appealing medical research frontier'.[44]

In the wider healthcare world, this was also a period of academic expansion and further specialisation in clinical services. It became common for leading figures to have both an academic role and be practising clinicians and a number of shared university and health service posts were created between regional centres and cytogenetics laboratories. Key centres developed in Aberdeen, Glasgow, Edinburgh, Newcastle, Manchester, Yorkshire, Birmingham, Liverpool, Oxford, Cambridge, London, Wessex, and Wales – the characteristics and idiosyncrasies of each of which been well described by Harper.[45] Bobrow recalls that the Oxford MRC unit was 'built in the grounds of a hospital trying to interface with the 1970s NHS. I suspect that it was a bit of a laboratory for both the MRC and the NHS. It wasn't unique, there were other places like that, but it was an interesting breeding ground for trying to work out how the interface between research, genetic service and clinical practice would work'.[46] Medical genetics was emerging as a distinct, functioning community, therefore, but progress was still slow. As Harper recounts, 'developments in Britain during the 1960s and 1970s were mostly piecemeal, ad hoc, financially modest ... the pattern (insofar as any pattern existed) was incre-mental development to occur around a single original founder'. Harper also noted the independence of centres, such as the one he helped to develop in Cardiff: 'this had nothing to do with official health policies, had nothing to do with [the Department of Health]'.[47]

These kinds of dynamics were perhaps most apparent in relation to sickle cell disease, thalassaemia and other haemoglobinopathies. In part, such genetic con-ditions did not become significant policy issues because they disproportionately affected ethnic minority groups. As the historian Roberta Bivins has described, 'Simply put, until the late 1960s, despite their scientific interest, the haemoglo-binopathies offered mainstream politicians little useful purchase on current affairs, while ignoring these obscure conditions presented no risk of public criticism'.[48] Key figures such as David Weatherall and other members of the 'biomedical elite' were concerned about the organisation of services and treatment and care, but they also recognised that increased funding for research would help to advance the field of medical genetics and maintain British scientific prestige.[49] Clinical services did not receive as much financial support, with moves to address this sometimes becoming caught up in wider organisational problems inside the Department of Health and Social Services and uncertainty about where responsibility for new service funding lay between the Department and the MRC. As Bivins describes, 'DHSS refused to fund Modell's work [on molecular prenatal diagnosis] on a service basis. No matter how valuable such diagnostic techniques were proving to be for affected families, from the Department's perspective their development merely constituted long-term research by clinicians'.[50] Services for conditions such as thalassaemia and other haemoglobinopathies developed incrementally around regional centres, drawing on a patchwork of local and central funding. A crucial role in ensuring better provision was played by community activists and pioneers such as Elizabeth Anionwu and Milica Brozovitch who established the first Sickle Cell and Thalassemia Centre in Brent in North London in 1979, which brought together screening, counselling, and treatment services.[51]

From 1973 recombinant DNA technology – a form of genetic engineering in which elements of DNA from different organisms are cut and pasted together – allowed the formation of new sequences and facilitated further discoveries. From 1977, the first scalable, economic method for reading the DNA sequence – Sanger Sequencing – was developed by the British biochemist Frederick Sanger. According to Malcolm Ferguson-Smith (another colleague of McKusick who led the development of medical genetics in Glasgow before moving to Cambridge in the late 1980s): 'cytogeneticists have always been slaves to technology'.[52] The faster and less expensive technique meant that a reliable network of cytogenetic services could be developed across the country, and genetics increasingly came to have a well-established place in British medical practice. According to Paul Polani, 'Medicine too had to come to grips with the practical uses of cytogenetics and genetics in the clinic, and so came to realise its fundamental contribution to basic science, level with anatomy and physiology'.[53] In 1980, clinical genetics was officially recognised and added to the Department of Health and Social Services' list of medical specialties for the first time. This was due in large part to the work of Cedric Carter – long-time Director of the Clinical Genetics Unit at the Institute of Child Health in London – who had established the Clinical Genetics Society (CGS) in 1970.[54] The CGS brought together medical professionals with an interest in genetic disorders and facilitated discussions about differences in the provision of tests, and the role of the geneticist, their workload, and their training and accreditation. The CGS, and Carter's appointment as the first Consultant Advisor to the Chief Medical Officer on Genetics in 1972, are considered in more detail below.

The 1980s was a further intense period of scientific breakthroughs. In 1983, the first human disease gene – Huntingdon's disease – was successfully mapped by the Canadian geneticist James F. Gusella and colleagues, while in 1984 the Polymerase Chain Reaction (PCR) method was developed in the United States, allowing small sections of DNA to be amplified for study. These technological advances underpinned a new phase in which it became possible to analyse at a molecular level how genes influenced the risk of disease. The specific gene mutation responsible for Duchenne muscular dystrophy was identified in 1986, and for cystic fibrosis in 1989. This evolution from cytogenetics to molecular genetics was perhaps best symbolised by the replacement of the microscope by the computer as the main analytical tool.[55] These new diagnostic capabilities also aided the development of a distinct community of genetic counsellors.

Yet genetic services continued to be built from the ground up in a relatively informal way. Where wider structural and institutional support did develop it largely came from Regional Health Authorities, an important part of the NHS landscape between 1974 and 1996, which organised resources and planned services in local areas, wielding significant power and autonomy.[56] Nineteen regional centres were established with genetic laboratory services and counselling clinics, as well as satellite clinics at local District General Hospitals.[57] This dynamic has been studied by Peter Coventry and John Pickstone who highlighted the ability of consultant geneticists in Manchester to integrate research, laboratory work and clinical services in a way which was appealing and fitted with the 'hospital dominated political economy of medicine'.[58] Support for medical genetics also came from the Royal College of Physicians (RCP) which established a Clinical Genetics Committee in 1984 (at a time when Cyril Clarke, who had been RCP President, became director of its Research Unit). It produced a series of reports on screening, teaching, and ethics, which were influential in establishing what should happen or what services should be provided in population terms at a regional level.[59] The Committee also acted as a conduit through which the interests of practitioners could be fed into the health policy system and was effective in drawing extra funding into genetic services.[60] As Peter Harper explains:

> One could put forward one's ideas … ideas of the specialty, through the College of Physicians, which would then send them on to the Health Department and the regions, and then six months later one would get a sort of memo from whatever Authority one was with, saying 'the Royal College of Physicians has issued a report on this, this and this, and it looks really important, will you sort of follow those lines?' and of course they didn't realise that those were the lines we had written ourselves![61]

Harper was a member of the working group, alongside Bobrow, Weatherall and others, which produced the report *Prenatal Diagnosis and Genetic Screening: Community and Service Implications* in 1989, which argued the case for such services, calling for more resources, better education and training, and a new policy advisory structure.[62] Even so, the Department of Health was still less than fully receptive to medical genetics during the 1980s – an issue that is considered in more detail below. Central planning appeared to Harper to be 'non-existent', but again this was not necessarily an issue. According to Harper: 'to be quite frank with you that was a very good thing, because what we have seen ever since is a whole series of mostly misguided and unrealistic initiatives'.[63]

By the mid to late 1980s, regional genetics services, which brought university research and services in major hospitals together, were well established. John Burn describes the development of the Northern Genetics Service as 'very much a graft on to the side of a hospital in a little terraced house, and we started taking over more and more space in our vicinity and metastasizing'.[64] Burn became Newcastle's first NHS consultant in clinical genetics in 1984 and its first Chair of clinical genetics in 1991. A critical mass of expertise had been built up which, supported by collaboration between the leading centres, stimulated new genetic specialisms in cancer, cardiology, ophthalmology, psychiatry, and other areas.[65] Cancer in particular was important as a common condition that could be increasingly well understood in terms of Mendelian inheritance. Clinical genetics broadened out substantially and was increasingly well-integrated into mainstream health care. The field had developed from one based on individual action and 'personal specialisation' to one based on distinct groups and 'task specificity'.[66]

No sooner had these gains been made than they were undermined by a change in broader NHS policy. Following the 1989 White Paper *Working for Patients*, the 1991 'internal market' sought to introduce elements of managed competition into the NHS, differentiating between 'purchasers' who would commission services and 'providers' who would deliver them. Leading geneticists, who had built up their networks of centres and services through cooperation, found these service changes disruptive. It stymied plans for a national consortium which sought to build on developments in Scotland led by Rosalind Skinner of the Scottish Home and Health Department. Skinner had originally qualified in medicine before joining the Human Genetics Department at the University of Edinburgh in 1972. Seeing the potential overlaps around disease prevention, she later retrained in public health and, after a spell as a Registrar with the Lothian Health Board, decided to try and shape policy from the centre.[67] A Scottish consortium had been organised by the leading geneticists in Glasgow, Edinburgh, Aberdeen, and Dundee, with responsibility for the provision of different genetics tests shared. Having initially been supported by temporary research funding, Skinner secured central funding for the consortium in 1988.[68] Similar attempts to address the effects of duplication and the small number of tests being carried out for some rare genetic conditions across the whole of Britain however, remained undeveloped. According to Peter Harper:

> When [these plans] were knocked on the head it caused proportionately more damage, but luckily by the time it happened medical genetics had developed very strong roots. It hung together well between the different regions, at a clinical level, it had also developed strong links between the clinical and the lab level, and so that proved very difficult for the Department of Health to destroy, although I have little doubt that if it had been easier they would have. It meant there were several years of hanging in there for survival, at a time when we could have been racing ahead, far ahead of most other countries.[69]

It would be some years before real coordination and cooperation was possible again.

The Arrival of Genomics

Alongside these developments in clinical genetics, by the mid to late 1980s, a number of scientists were beginning to look at the potential of genomics. While genetics was concerned with the study of genes and their role in heredity, genomics sought to take account of the whole genome – all of an individuals' genes, all the DNA inside each cell – and the interactions between those genes. It had been possible to understand the genetic contribution to an ever-increasing number of single-gene disorders, but new technologies now presented the possibility of also being able to understand the genetic contribution to common complex conditions. A confluence of research in genetics and molecular biology had provided advances in the form of recombinant DNA, sequencing, and gene mapping. New techniques including Fluorescence In Situ Hybridisation – the use of fluorescent probes which stick to specific places on the human chromosomes in order to see if they are present, absent, or duplicated – allowed more disease-related genes to be isolated. It was first used clinically in 1992. The first full sequencing of a genome – the bacterium *Haemophilus Influenzae* – was achieved in 1995. From 1997, array CGH (comparative genomic hybridization) testing – through which gains and losses can be ascertained in any part of the genome – was available. From 2008, Next Generation Sequencing became possible, allowing whole genomes to be sequenced cheaply and quickly.

Such advances were underpinned initially by the wider race to sequence the whole human genome, headlined by the international Human Genome Project (HGP) which had been launched in 1989. The HGP has variously been described as 'arguably the most important event in biology since modern medicine began' and 'as influential as Darwin's theory of evolution'.[70]

The initial impetus for the project came from the United States, with the main aims first articulated by a special committee of the National Academy of Sciences in 1988 and then implemented through a partnership between the National Institutes of Health and the Department of Energy. The central premise was that sequencing the whole human genome was an important endeavour and that a more systematic and coordinated approach was needed to achieve it, moving beyond focussing on a particular disease or chromosome region as was then usual in most genetics research.[71] James Watson, then Director of the Cold Spring Harbor Laboratory in New York, was appointed as the first Director. It was apparent to all involved that the scale of the project and the sequencing capacity that was required necessitated a truly international collaboration. Research centres in the United States, Britain, France, Germany, Japan, and China joined the initiative, with hundreds of sequencing machines continuously processing DNA samples.[72] The final cost of the project – which was eventually completed in 2003 – was around $3 billion.

In Britain, a number of individuals had been making the case for sequencing the entire human genome. One influential figure was the South African biologist Sydney Brenner, Director of the MRC Laboratory of Molecular Biology at the University of Cambridge, and a future Nobel Prize winner for his work on the genetic regulation of organ development. He and others began to argue the case at meetings of the MRC, drawing on research using nematode worms. David Bentley, then a researcher at Guy's and St Thomas' Hospital and subsequently a leading figure in the development of the HGP in Britain, recalls of one MRC meeting: 'the ideas were really very open ... It was still very visionary, the whole field.

It was not well formed, but very exciting to hear'.[73] How this vision might actually be implemented was a difficult question. Keith Peters, Regius Professor of Medicine at Cambridge between 1987 and 2005 describes how 'the aeroplane was going down the runway quite slowly' until American interest demonstrated that something significant was going to happen. Peters was influential in bringing the HGP to wider attention, and ultimately in persuading the Prime Minister, Margaret Thatcher, that Britain should play a role. He was a member of the Advisory Council on Science and Technology (ACOST) established in 1987 in order to bring science closer to government. Convinced by Brenner's arguments, he pushed for the HGP, presenting it as not just scientifically but economically ambitious. Thatcher, herself a trained chemist, 'got it'.[74] Peters recalls:

> The argument at the time was very simplistic, you had a kind of blueprint, like an architectural plan of the human body … and you could work out what the genes were actually doing … that turned out actually to be hugely oversimplified, but at the time it seemed like an easy way of presenting it.[75]

The potential for better understanding and then treating cancer was an important part of the case for the HGP. Sir Walter Bodmer, Director of Research at the Imperial Cancer Research Fund, was a keen advocate. His career had spanned both UK and US institutions, and he had been a pioneer in establishing population genetics as a distinct area of study. He presented alongside Brenner at an ACOST subcommittee.[76]

Although the scientific merits of the HGP were accepted, the issues of practicality and cost were more difficult to resolve. Scientists drew on Moore's Law in IT – that the computing power of microchips doubles roughly every two years while the cost halves – to argue that sequencing costs would fall as the technology was developed.[77] An initial investment of £11 million over three years was made by the British government through the MRC and the Wellcome Trust. The Sanger Centre was established as a commercial vehicle in Cambridge in 1993, becoming one of the four key international sequencing centres. It was a strategic move that allowed relevant scientific knowledge to be retained in Britain, and technological capacity, particularly biotechnological capacity, to be built in order to underpin future developments. The HGP was an early example of Britain's new approach to commercialising scientific research, though some in the scientific community had concerns. According to Bobrow:

> Any sort of issues around simply 'we shouldn't be going too fast on this area', all the ethical concerns that were raised around cloning and stem cells and gene therapy and so on were seen as a bit of drag on this amazing economic thing.[78]

The extent to which the ethical, legal, and social implications of the HGP were considered in the United States and the impact on the early development of public health genetics are considered in more detail in Chapter 2. It was clear that the HGP would have significant implications for medicine and health, but also that they would take time to develop. Healthcare innovations should not be expected overnight. Indeed, for most of the 1990s clinical geneticists in Britain operated in a 'separate world'. As Burn reflects:

> We were finding genes anyway, you know. The Mendelian search for genes was happening separate to the Human Genome Project, which initially was just a whole bunch of Yanks linking old bits of DNA together. It was a bunch of boffins, in a way. In a sense they were sort of Bletchley Park to us being the Marines … We were out fighting, and we knew those guys were cracking codes and stuff, which was great, but we didn't really … get that involved with them.[79]

The early development of public health genetics during this period, principally through Muin Khoury in the United States and Ron Zimmern in Britain – who are discussed in more detail below – rested on an anticipation that public health theory and practice would need to evolve in response to genomic knowledge, but also that there would be a need to 'counter the hype'.[80] According to Zimmern, '[I] understood that it would revolutionise biology, and I suppose my question was what the hell are policymakers and the public health tribe going to do about that?'[81] This wider shift from genetics to genomics was at an early stage, but it was important to the emergence of public health genetics. In order to fully understand this emergence, we need to consider the wider context of the development of public health.

The Development of Public Health

Many of the individuals that shaped the field of public health genetics were initially trained in and practiced public health medicine. These experiences often instilled a set of values and a perspective that stayed with them, founded on an appreciation of the value of population-wide approaches to health and healthcare-related questions. This was often married to an interest in the possibilities of genetic and genomic medicine but not subsumed by it. Public health – as a discipline and practice – has a nuanced history that shares many of the same determinants as genetics and genomics: variable professional and political support, difficulties establishing academic centres of expertise, chronic funding problems, and the challenges of presenting complex ideas to the public and implementing policy changes after scientific advances. The history of public health includes periods of expansion, revision, retraction, and reversion. While infectious epidemic diseases are no longer as prevalent as they once were, and current generations are living longer than their predecessors, the overall story is not one of simple progress A central feature of public health has been its broad basis and flexibility. This means that it has been inclusive and responsive to need, but also that at important moments it has lacked cohesiveness. The difficulty of defining 'public health' has been both its strength and its weakness.[82] As the historian Christopher Hamlin neatly puts it: 'The great debate in the history of public health is what public health is and should be'.[83]

As with genetics, there is evidence of public health philosophy and practice in antiquity. Ancient Greek physicians such as Hippocrates and Galen were known to have stressed the importance of hygiene.[84] During the Middle Ages and the early-modern period there were sporadic but often unsuccessful attempts to stop the spread of destructive epidemic diseases. From the late eighteenth century, the Industrial Revolution stimulated rapid and unplanned growth in many British cities. Overcrowding and lack of basic sanitation such as piped water and sewer systems meant that infections spread rapidly in urban areas. This happened at a time when local government received limited funding and was restricted in the actions it could take. It was not thought to be the role of the state to interfere in the lives of individuals. The underlying philosophy was one of 'laissez faire'. Even though theories of disease transmission centred on 'miasmas' or 'bad air', from the 1830s there was increasing evidence of the association between ill-health and poor living conditions. Pioneering social reformers such as Edwin Chadwick led surveys that demonstrated a clear link between poverty and life expectancy, and identified the wider costs to the whole community for the first time.[85]

Public health is at its heart a 'local' science. The first practitioners to identify themselves with the discipline were individuals with an excellent knowledge of their local communities. During the mid-nineteenth century, the places with the worst health outcomes therefore produced some of the best experts. In Liverpool – known at the time as the 'Black Spot on the Mersey' because it had some of the highest mortality rates in Britain – a key group of reformers carried out detailed street-level analysis to supply evidence to Chadwick's 1842 national inquiry. The most pro-minent member was the Liverpool-born physician, Dr William Henry Duncan.[86] His reports formed the basis on which the town successfully petitioned parliament for a local 'Sanitary Act' in 1846. This allowed the town council to create three world-first posts: Medical Officer of Health (MOH), Borough Engineer, and Inspector of Nuisances. Duncan was appointed as Britain's first MOH, an innovation subsequently adopted by other towns and cities under the 1848 national Public Health Act. A raft of mid-nineteenth-century legislation facilitated a range of sanitary reforms, including the municipalisation of water supplies, building sewer systems, isolation hospitals for infectious diseases, public baths and washhouses, and standardised housing quality – all funded through local taxation. At the national level, following a series of devastating cholera epidemics, the government recognised the need for greater oversight of public health. It created the post of Chief Medical Officer (CMO) in 1855 and appointed the epidemiologist Sir John Simon to the role. Although the resources under Simon's control were minimal (he had no line management of local MOHs), he and subsequent CMOs were able to build up a team of medical civil servants who could be deployed when infections flared up around the country, and who could conduct studies and advise on how to improve public health.[87]

Epidemiology – the science of studying causes and patterns of disease within populations – has long been one of the central components of public health practice. Duncan may not have described himself as such, but he was clearly as skilled an epidemiologist as the recognised 'founding father' John Snow, who earned a living as a physician and pioneering anaesthetist.[88] Snow's detailed analysis of the distribution of cholera cases around a water pump in Broad Street, London in 1854 demonstrated that the disease was not caused by 'miasmas' but was water-borne. This was such a radical, indeed heretical, break with the established understanding of disease transmission that had lasted for centuries that it was not fully accepted by the scientific community during his lifetime. Snow died in 1858. Despite his breakthrough, the science and practice of public health remained essentially 'sanitary' until the late nineteenth century when scientists such as Louis Pasteur and Joseph Lister contributed to the emergence of a distinct 'germ theory' of disease – from the Latin for 'seed'. This was an all-encompassing term that included more specific discoveries of the organisms that caused individual infections, such as the cholera vibrio and the tuberculosis bacterium identified by the German microbiologist Robert Koch during the 1880s. These breakthroughs provided the foundations for the science of bacteriology to become a partner to epidemiology in the public health practitioners' toolkit. Now individuals as well as food and water could be tested for invisible infections. This transformed the possibilities of public health, at a time when medicine was still relatively limited in its therapeutic responses to disease. Prevention was now a real possibility, especially with the creation of vaccines for diseases such as smallpox and diphtheria.

With these new tools available, public health practitioners became increasingly powerful professionals within their local authorities, especially those that worked as MOHs for large city councils. They could condemn slum housing, shut down unsafe food sources, inspect children, and isolate individuals with notifiable infections. All of this was possible despite the inability to treat many infections or address the underlying determinants of most ill-health – which were now well-known to be linked to poverty.

By the early twentieth century, there was a clear epidemiological transition in Britain. Epidemics of infectious diseases such as cholera and typhus had subsided and been replaced as leading causes of death by chronic conditions, especially coronary heart disease and cancer, although pandemics continued to appear. The 1918–1919 influenza pandemic was particularly devastating, causing between 30 and 60 million deaths worldwide.[89] When the influenza pandemic hit Britain, the national response was much the same as it had been during nineteenth century cholera outbreaks – managed by a public health service dislocated between MOHs in local authorities and the CMO at a national level at the Local Government Board. However, the state's responsibilities for public health, if not for personal health services, now extended into areas such as vaccination programmes, infant and maternal welfare centres, and sexual health initiatives. A fundamental restructure was required to provide clear political responsibility.

In 1919, the Local Government Board was disbanded, and the health functions moved to a new Ministry of Health. Further reforms in 1929 gave MOHs local responsibility for hospitals that had previously been created as Poor Law infirmaries. They were now consolidated into a municipal healthcare system, alongside infectious disease hospitals that had expanded their remit into general medicine. These rivalled their private 'voluntary' counterparts in terms of bed numbers, staff, and treatments, if not in terms of prestige.

MOHs – all medically qualified from Duncan's appointment until the 1990s – organised themselves into a formidable professional community during the inter-war period, with considerable resources at their command. They were passionate about the health of their local populations and experts in understanding its determinants. Notable MOHs included George McGonigle in Stockton on Tees, who demonstrated that poor diets were not a choice, but a result of rising rents and other living costs. In Liverpool, Edward Hope focused on persistently high levels of infant mortality caused by feeding milk from unsterilised bottles. However, he failed to identify that bottle-feeding itself was determined by the lack of maternity benefits, meaning mothers had to return to work immediately after childbirth. Public health was also becoming more politicised, as demonstrated by the scandal around diphtheria vaccinations, which the Canadian public health authorities had successfully introduced during the 1920s. The British government was well-aware of this but chose not to start its own scheme until 1944 because it was not seen as a financial priority, despite evidence that around 6,000 children died each year from the preventable infection.[90]

The creation of the National Health Service in 1948 provided a golden but lost opportunity to fully integrate preventive and curative medical services. The 'tripartite' system introduced by Health Minister Aneurin Bevan kept public health services and staff within local government, with only a loose association with GPs and hospitals. The failure to build a planned series of linking health centres exacerbated the dislocation, with most GPs remaining as 'single handed' practitioners, often working from their homes. Even so, public health as a discipline and practice continued to evolve from the 1950s, with a new focus on diseases of lifestyle and behaviour, stimulated by studies of the causal relationship between smoking and lung cancer, led by Richard Doll and Austin Bradford-Hill, and studies on physical exercise and coronary heart disease, led by Jerry Morris. Epidemiological research, whether led by MRC-funded units or traditional academic departments of public health, supported the development of effective health promotion strategies, and was championed by a succession of high-profile CMOs, the most outstanding of whom was George Godber, between 1960 and 1973.[91]

The broad church of 'public health' has always invited collaborations with disciplines and methodologies beyond the conventional epidemiological foundations. The longstanding recognition that much ill-health and health inequalities were linked to poverty stimulated fruitful partnerships, with sociologists, economists, and experts in social administration. During the 1950s, Richard Titmuss had argued for the state to take a leading role in providing health and welfare in order to achieve 'social justice'. He worked alongside Brian Abel-Smith and Peter Townsend, who developed the influential concept of 'relative poverty'. New data sources – from household surveys to birth cohort studies – and the arrival of computers in both universities and NHS services, were important from the 1960s. The pioneering Oxford Record Linkage Study opened up new possibilities for synthesising interests in public health and health services. When public health moved into local health authorities after the 1974 NHS reorganisation (the first significant set of changes since its creation in 1948), practitioners – now re-labelled as Community Health Physicians – had the data and the tools for effective population health surveillance. There was, however, a sense that the role had lost some of its enabling functions, given that its connections with local government had been cut.

By the 1970s and 1980s, more public health academics and practitioners were willing to be seen as political advocates. Building on the 1940s tradition of 'social medicine' followed in the 1960s by the concept of 'community medicine', demands to address the poverty that underpinned health inequalities became stronger. In 1977 the then Labour government established a commission, led by Sir Douglas Black, the government's chief scientist, to examine health inequalities. Jerry Morris was a member of the inquiry team. However, when the Black Report was ready in 1980 it was delivered to a new Conservative government. It swiftly rejected the report's recommendations for addressing health inequalities, despite persuasive evidence that the poorest communities were failing to benefit from the free-at-point-of-delivery NHS, and that by investing in public health and reducing poverty the now spiralling costs of health care might be brought under control. A significant reallocation of resources was not on the agenda. Only two hundred and sixty copies of the report were published, and it was released over the August bank holiday weekend.

No matter how public health academics presented their robust, evidence-based findings, such as Margaret Whitehead's 1987 book *The Health Divide*, and Richard Wilkinson's 1996 book *Unhealthy Societies: the Afflictions of Inequalities*, there was a political impasse on policy responses.[92] The government even went as far as to ban the use of the term 'health inequalities', requiring that all official reports use the less contentious phrase 'health variations'. Staff and budgets in local authority departments were cut, and outbreaks of rare and novel infections such as legionnaire's disease and HIV/AIDS highlighted the hollowing out of public health capacity at both local and national levels. Yet, the public health community continued to draw strength from long-established collaborations, especially with environmental health and social services, finding new ways to address challenges through the creation of multi-disciplinary public health teams. Directors of Public Health – as Community Physicians became in 1982 – no longer had to be medically qualified, and the Faculty of Public Health reframed its relationship with the Royal College of Physicians, dropping 'medicine' from its title.

The 1988 Acheson Report – produced by Sir Donald Acheson, CMO between 1983 and 1991 – into the future role and development of public health defined it as 'the science and art of preventing disease, prolonging life and promoting health through organised efforts of society' and called for it to have a more central role in debates about health and health care.[93] This approach resonated with practitioners. But it was to take until the late 1990s for public health to receive much greater political recognition. This was linked to the new focus on the 'social determinants of health', reflecting the shared approaches of public health academics and professionals, and the acknowledgement that health is linked to factors such as income, housing, and educational levels, and that these in turn shape individual choices on key issues such as diet, exercise, smoking and alcohol intake. The social determinants of health approach also underpinned health management strategies, including targeted screening invitations screening for conditions such as cancer and heart disease.

Many of the early initiatives associated with public health genetics therefore came after a period in which public health had been in the relative political wilderness in Britain, but also one in which many academics and practitioners were eager for new challenges and collaborations. It was a relatively simple conceptual step to add 'the application of advances in genetics' to Acheson's definition of how public health could be achieved, emphasising the longstanding recognition that gene-environment interactions were crucial health determinants. As Mikail has described, 'Genetics and core public health concerns continued to intersect, and their shared territory continued to grow'.[94] Genetics and public health came together to ensure that the promise of the former was successfully realised As Hamlin has described, 'History is frequently deployed in the shaping of public health institutions … One recognizes ways in which a society has not facilitated health and explores how it might better do so. There are no *a priori* boundaries to the changes that might be promulgated in pursuit of health'.[95]

Genetics and Public Health

While public health genetics began to crystalise during the 1990s as the possibilities of genetic and genomic medicine became clearer, there had been important intersections between genetics and public health before, which speak to the compatibility of the two fields. For example, there were shared interests in and around newborn screening programmes from the mid-twentieth century onwards. Phenylketonuria – a metabolic disorder associated with intellectual disability in which damaging levels of the amino acid phenylalanine build up in the brain – was first identified by the Norwegian physician Asbjörn Fölling in 1934.[96] Its significance as a genetic disorder, as well as potential therapies, were discussed by Lionel Penrose during the late 1940s, and the efficacy of low phenylalanine diets as a treatment was demonstrated by the German paediatrician Horst Bickel in 1953.[97] The potential for such treatment and advances in cytogenetics facilitated prenatal and neonatal diagnosis of a number of chromosomal disorders like phenylketonuria, and by the 1960s urine samples were regularly taken from newborn infants by British health visitors as part of population-level screening to detect and prevent the condition. In the United States, with notable legislative support, most states had developed detailed Phenylketonuria testing programmes by the mid-1960s.[98] As such, there was an appreciable confluence of interests around genetic detection and disease prevention at a population level. As Muin Khoury has noted, 'At that time the words "public health genetics" didn't even exist' but 'new-born screening is the ultimate public health genetics programme'.[99]

The lack of formal training or professional recognition for geneticists, and the ad-hoc way in which services developed, meant that genetics was often put into practice by those with other medical training. As William Leeming notes:

> Screening services [in Britain] were provided throughout the country at university children's' hospitals by paediatricians, obstetricians and other specialists in the field of neonatal medicine, they remained managerially and operationally separate from the type of services associated, years later, with genetic diagnostics and counselling in regional genetic centres. Likewise, the role of specialist geneticists in the diagnosis and management of Rh incompatibility, the thalassaemias, and other haematological conditions remained marginal in relation to that of haematologists, obstetricians and specialists in blood transfusion and serology.[100]

This dynamic meant that consideration of issues relevant to public health was often embedded in genetics activity. For example, although many leading geneticists passed through the Oxford MRC unit and it did much to help develop clinical genetics as a whole, Alan Stevenson, its Director between 1958 and 1974, studied congenital malformations and inherited diseases such as muscular dystrophy and cystic fibrosis as population health issues – as Cedric Carter had done in London drawing on his background in medicine and public health.[101] Although initiatives such as screening programmes demonstrated the potential of genetics, they also did so without undermining the possible importance of other factors such as environment. Genetics and public health could be complementary. Susan Lindee has described how genetic testing was central to phenylketonuria management, but because the appropriate treatment was dietary control it was clear that 'biology is not destiny'.[102]

While some clinical specialisms took longer to appreciate the potential value of genetics, the study of gene-environment interactions increasingly took a central place in conceptions of disease aetiology. A series of genetic sub-specialties emerged, often with a related public health approach built in. For example, cardiac genetics sought to address the under-diagnosis of genetic disorders such as hypercholesterolaemia, which fell between the cracks around cardiological and genetic services. In the 1990s, genetic screening programmes were developed to exploit the discovery of genes associated with cardiomyopathies and cardiac channelopathies.[103] Screening programmes for conditions such as Down's syndrome were established after the development of prenatal genetic tests, though the ethical implications of thinking about such tests in public health terms have been the subject of significant debate.[104]

The complementary nature of genetics and public health manifested itself most clearly however in the field of genetic epidemiology. The term was originally coined by the geneticists Newton Morton and Chin Sik Chung during the 1970s to describe a developing science 'which deals with the aetiology, distribution, and control of disease in groups of relatives and with inherited causes of disease in populations'.[105] It drew on the longstanding traditions of epidemiology and sought to incorporate new genetic and statistical approaches. Muin Khoury, who began a PhD in genetic epidemiology at Johns Hopkins University School of Public Health in 1982 – the first formal programme of its kind which began in 1979 – recalls how he wanted to 'understand why some people get sick and others do not, even when exposed to the same environmental insults such as smoking and infectious agents'.[106]

Khoury was mentored by Bernice Cohen, who had laid much of the groundwork in bringing the two fields together, building on studies from the 1950s which sought to bring concepts from genetics into epidemiology. Cohen had been a 'disciple' of Abraham Lillenfeld, a 'giant' of epidemiology at Johns Hopkins who had broadened the field to study chronic as well as infectious diseases. Alongside their colleague Terri Beaty, Cohen and Khoury produced the influential text *Fundamentals in Genetic Epidemiology* in 1993. The book described the bringing together of the genetics and epidemiology over several decades as a 'gradual rapprochement'.[107] By then Khoury was working on birth defects and congenital anomalies at the Centers for Disease Control in Atlanta, but he remained aware of the importance of having this broader perspective. As he recalls, 'the progress was exciting to me ... at CDC I kept making noise'. This noise was fundamental to the subsequent development of public health genetics. The establishment of the CDC's Office for Genomics and Public Health in 1997 is discussed in Chapter 2, and the significance of Khoury's international collaborations is considered in Chapter 4.

By the mid-1990s, Khoury and others could see that the rapid evolution of genomics was likely to provide a fundamental breakpoint. Genetic epidemiology developed a focus on technology rather than theory, underpinned by better understanding of disease at a molecular level, computational statistics, and significant advances in genome sequencing.[108] The field has now been widely integrated into clinical and public health medicine and speaks to issues around personalised medicine and pharmacogenetics. Yet these developments stimulated calls for further philosophical debate on the potential implications of genomic medicine. While genetic epidemiology by its nature focussed largely on research, public health genetics sought to fill this gap, confronting difficult ethical and legal issues as well as important practical questions, trying to ensure the widest possible population health benefits from genetic and genomic medicine. The relationship between genetics and public health subsequently became more formal and systematic, transitioning from screening programmes for single-gene disorders and inherited conditions to genomic medicine as a whole, with much broader implications. Public health genetics may not have been pushing at an open door, but the door had been unlocked and oil had recently been applied to the hinges by a series of leading researchers and practitioners.

Official Interest in Genetics

As medical genetics developed it naturally came to the attention of civil servants, politicians, and other health policymakers, moving slowly up the agenda from the 1970s on. An early sign of it doing so was the appointment of a Consultant Adviser to the Chief Medical Officer on Genetics in 1972. The CMO, at the time the influential George Godber, drew on the knowledge and experience of a panel of eminent physicians from across the medical specialties. This was part of a hierarchy of advisory committees under the umbrella of the Central Health Services Council that were supported by the Department of Health and Social Security to provide advice to the Secretary of State for Health.[109]

Securing representation through the appointment of a Consultant Adviser was a sign that a clinical specialty had matured.[110] For context, in 1964 there had been 39 consultant advisers covering 30 specialities, with another for general practice.[111] By 1976, there were 49, and by 1991 there were more than 100.[112] The first to take on the role for genetics was Cedric Carter. Carter had established the Clinical Genetics Society in 1970 and was influential in pushing for medical genetics to be recognised as a specialty. As Director of the Clinical Genetics Unit at the Institute of Child Health, he was also central to the development of the field on the ground. Through its close links with Great Ormond Street Hospital, the Unit became a focal point for the study of childhood syndromes and malformations and laid many of the foundations for genetic counselling. A number of future 'stars of the field' such as Marcus Pembrey, Michael Baraitser, Robin Winter, and John Burn also passed through the Unit.[113] Beyond his leadership in the field, however, Peter Harper speculated that there may have been some more hard-headed thinking behind Carter's involvement with DHSS:

> One wonders also if his support for eugenics, virtually alone in this among the post-war medical genetics community, may have been an unspoken factor in the minds of those health policy makers who saw the 'prevention' of genetic disorders mainly in financial terms.[114]

Although there was official interest in genetics it is important not to overstate its significance at this early stage. In 1972 the Standing Medical Advisory Committee – another independent advisory body in place since 1949 – produced a slim pamphlet titled *Human Genetics*, ostensibly for the Central Health Services Council and the Secretaries of State for Social Services, Scotland, and Wales, but with the aim of providing doctors with a better understanding of genetic disease. The mechanisms of dominant, recessive, and X-linked inheritance, and chromosomal abnormalities, were described, as well as the overall risks of different conditions developing in different situations. It was recognised that genetic factors were likely to have a role in common complex conditions, but there was as yet no real sense in which the relevant aetiology might be understood beyond observations of familial concentration, which might often be the result of coincidence.

There was no suggestion that it was the responsibility of policymakers at the centre to facilitate discussions about the development of services.[115] When Paul Polani was later asked by Harper to name a civil servant with a direct interest in genetics he was unable to do so, and Ian Lister Cheese, who joined DHSS in 1983 and would become an important figure, recalled how 'I read into the subject, as one must; there wasn't very much within the Department … there was very little by way of a policy statement, very little indeed'.[116] For the moment, the running was still being made by the geneticists themselves, taking advantage of health service funds where available, developing their own polices and building research and service capabilities. According to Lister Cheese:

> It didn't matter, because the work of the professional bodies, the RCP, the Clinical Genetics Society, and of individuals – many of them – had led to a series of authoritative documents which laid down clearly what a clinical genetics service should comprise. They set out its purpose, its function, and all the elements that make up an effective clinical genetics service. They described its educational and research roles and they had investigated the ethical aspects of genetics. There was a sense in which one didn't have to do anything. The ground had been laid.[117]

When Carter retired in 1982 he was succeeded as Consultant Adviser by Rodney Harris, Professor of Medical Genetics at the University of Manchester, and Chair of the Committee on Medical Genetics of the Royal College of Physicians. Harris was a 'formidably adept politician' and had an established relationship with DHSS.[118] The broader development of medical genetics during the 1980s and the eventual integration of Regional Centres owed much to his 'vision' and 'influence'. Regional genetics services had been developed in several places, including Manchester, during the 1970s, but the pace of change was slow, and variation persisted. The Clinical Genetics Society (CGS) and others were anxious to go further. A report by a CGS Working Party in 1982 suggested that although DHSS had been aware of the need to better develop regional services, more central coordination and more resources were needed, particularly as advances in technology were likely to further drive up demand.[119] Harris also made this case and in 1984, following discussions with Ministers, DHSS supported a pilot study at three centres – Cardiff, Manchester and the Institute of Child Health in London – through which funding was provided to facilitate integration, including bringing DNA laboratories into the health service. This was achieved via a 'Special Medical Development' (SMD), a process which had been introduced in 1974 to evaluate and plan for important emerging NHS services.[120] Integration of genetics services around regional centres was shown to be the way forward and Harris regarded this as his main achievement:

I think the Department of Health began to take medical genetics seriously as a specialty. Cedric, bless him, had got recognition for consultant status but really there were so few of them and it was possible to go along and say there ought to be more.[121]

However, the impact of the SMD exercise was effectively limited to supporting the three existing centres, with minimal wider benefit. According to Lister Cheese, it 'didn't really establish substantive policy one iota'.[122] In 1992, Harris was also influential in establishing a national 'Confidential Enquiry' – a type of investigation into poor clinical practice – on counselling for genetic disorders by non-geneticists. This was carried out by the Royal College of Physicians for the Department of Health and increased awareness of the importance of genetics in healthcare among professionals and purchasers and providers. The impetus was once again provided by advances in DNA technology through which elements of genetics were increasingly spreading to other specialities. Audits were completed of services for Down's syndrome, neural tube defects, and multiple endocrine neoplasia type 2, and the enquiry identified inadequacies in training for non-geneticists and inconsistencies in the way that important genetic information was presented to patients. Harris's role as Consultant Adviser to the CMO was helping to push genetics up the policy agenda and led to clear improvements in services and practice. But there were limits to what he was able to do. When later asked by Harper what it meant to be Consultant Adviser, Harris replied 'Everything and nothing':

Harper: OK, what about the everything.
Harris: Everything was, I could go on bashing away at genetics. We had people there who were the most senior GPs in the College of GPs, very erudite physicians and all the surgeons, everybody. The absolute top brass of medicine were there, and me pushing genetics, nobody had ever heard of before.
Harper: So did it give you an open door to sort of give your views?
Harris: At least to make the point. Yes it did, no question about that.[123]

The fact that there was more traction for genetics inside DHSS was reflected in a senior civil servant taking a direct interest, in the form of the 'tremendously helpful' Ian Lister Cheese. As a former general practitioner and Secretary of the Standing Medical Advisory Committee, Lister Cheese was receptive to arguments about the importance of genetics and was well placed to act as a 'good intermediary'. However, genetics was only one part of his brief and there was limited capacity within DHSS to take on further responsibility and not much willingness to facilitate the development of services. Lister Cheese was clear that while genetics was being pushed forward by new technologies – 'the department was not driven at all by such advances – it had to be prodded into action'.[124] John Yates, later Professor of Medical Genetics at the University of Cambridge, who was part of efforts to produce more joined up thinking later described how, 'It was very frustrating because the message that we were getting from the Department of Health was that they didn't want to facilitate any sort of coordination'.[125] This apparent reticence on the part of the Department of Health was subsequently writ large in the form of the 1991 internal market, which, in seeking to promote elements of managed competition in the NHS, actively undermined much of the coordination in genetic services that had achieved by geneticists on the ground.

In 1989, Marcus Pembrey became Consultant Advisor to the CMO on Genetics. Pembrey also had an established relationship with the Department of Health, having led the Special Medical Development activity at the Institute of Child Health in London:

> We got to know Iain Lister Cheese very well … Iain Lister Cheese and I would go to dinner every month at Rules in London, a very fine restaurant it was in those days, and so it was just very easy for me, compared with those in Manchester and Cardiff, to hobnob with the Department of Health.[126]

Pembrey describes the role as 'good fun' and important in terms of being able to help shape the development of genetic services, not least because, 'the Department of Health, really didn't know very much about genetics at all, so we were given a free rein'.[127] Yet by the 1990s, with continuing advances in genetic science, there was wider interest in related ethical questions and Pembrey's advice was sought on a more regular basis – for example, in relation to consanguineous marriage and the prevalence of recessive disorders. Similarly, when a particular need arose, there were now means through which it was possible to provide a steer from the centre and senior civil servants became more confident in providing some direction. For example, in 1993 the CMO Kenneth Calman and the Chief Nursing Officer (CNO) Yvonne Moores wrote a joint 'Professional Letter' on 'Services for Genetic Disorders' to NHS staff. Such letters allowed them to address a specific topic outside of regular communications and get across important information to relevant health professionals in the field. In this case, guidance was issued to health authorities in the form of the report 'Population Needs and Genetic Services', which had been produced by the Genetic Interest Group.

Genetics, Ethics, and Patients

The ethical implications of genetic and genomic medicine have been central to the conception of public health genetics and the development of much genetics policy, and here too there is a history of critical thinking that can be drawn upon. Tension between advances in science and technology and established ways of thinking about issues such as confidentiality, consent and trust have long been inherent in medical practice and research, but genetics raised such issues in acute form.[128] Since the emergence of eugenics there has been serious reflection about what advances in genetics would mean for individuals and society. The real-world implications were significant from an early stage – for example, in 1907 the US

State of Indiana became the first place to legalise non-therapeutic eugenic sterilisation – and the legacy of such policies is still keenly felt today.[129]

Debates about abortion and the legislative efforts which ultimately brought about the 1967 Abortion Act in Britain did not emerge from genetics, but they did provide the context in which many genetic developments took place and had an impact on research and practice, particularly in relation to prenatal diagnosis. One effect of the Abortion Act – which permitted the termination of a pregnancy if a detected foetal abnormality indicated a substantial risk of serious physical or mental handicap – was to increase demand for testing for genetic conditions. Developments in amniocentesis made prenatal screening for Down's Syndrome, neural tube defects and other conditions more common during the 1960s and 1970s. Crucially, in developing molecular prenatal diagnosis for thalassaemia, key figures such as Bernadette Modell 'paid explicit attention to the trans-national communities to which their patients belonged, and drew upon them for resources when facing apathy or resistance in their home medical institutions and bureaucracies'.[130] Peter Harper suggests that despite much caution, the implications for high-risk families were such that 'most medical geneticists were or soon became aware of ... keenly felt attitudes and adopted the view that, where couple wished for prenatal diagnosis in such situations, it was right to offer it, to give support and to work towards making safer and more feasible'.[131] The approach of the clinical community to abortion was largely 'thoughtful' and responsible', in contrast to 'the sort of convulsions there were in America and some other countries'.[132] Nonetheless, to some extent genetics did become entangled with abortion in the popular and political imagination. According to Iain Lister Cheese, 'throughout those years in the DHSS, much of our attempt to do rather more in respect of the NHS and its support for clinical genetics was overshadowed by a general unease about abortion'.

The birth of the first baby as a result of In Vitro Fertilisation (IVF) – Louise Brown in 1978 – following the pioneering work of Patrick Steptoe and Robert Edwards in Britain, also raised awareness of advances in genetics and embryology and provoked new ethical and philosophical questions. In 1972, a committee led by Sir John Peel, former President of the Royal College of Obstetricians and Gynaecologists, had established a code of practice for the use of foetuses and foetal material in research. The guidance was updated after a review in 1989 led by John Polkinghorne, who would continue to have an important profile in and around advisory bodies during the 1990s and 2000s. Even so, there was increasing recognition of the need to regulate and ultimately to legislate. A key turning point came with the establishment of the Committee of Inquiry into Human Fertilisation and Embryology in 1982, chaired by the moral philosopher Dame Mary Warnock. It was tasked with considering the social, ethical, and legal implications of advances in science and medicine. The Committee's report, published in 1984, recommended licensing for IVF and set out a time limit of 14 days for the use of foetuses in research. Its recommendations underpinned the 1990 Human Fertilisation and Embryology Act, which established the Human Fertilisation and Embryology Authority in 1991. The Warnock inquiry was explicitly pluralistic, and the philosopher and ethicist Baroness O'Neill reflects that it was important that such issues were addressed gradually, discursively and by Parliament rather than by the Courts.[133] By the 1990s therefore, there was a strong ethical underpinning to many debates about genetics practice and research. The anthropologist Sarah Franklin identifies IVF as an important 'disruptive technology', and the Warnock Committee as having established the principles around the governance of new forms of bio-medical innovations.[134]

The introduction of the Human Fertilisation and Embryology Bill in 1988 also led to the formation of the influential Genetic Interest Group. Provisions in the Bill that allowed for the use of human embryos in research under tightly controlled circumstances were opposed by a number of religious and 'pro-life' groups. In response, a range of patient organisations and representatives of families with in-herited genetic diseases coordinated their efforts, initially in a time-limited cam-paign, to make the case for the Bill. On the back of their success, a formal charity was formed in 1990. The first Director of the Genetic Interest Group (renamed Genetic Alliance UK in 2010) was Alistair Kent, who had worked for a number of disability charities and with school leavers with special educational needs as part of local government careers services. Kent saw genetics as having arrived at an im-portant moment, one in which it would be crucial that the voices of patients and their families were heard. As he describes, 'yesterday's science fiction was today's *Nature* paper, was tomorrow's potential clinical service improvement'.[135]

The Generic Interest Group was initially a very small umbrella organisation, with Kent as the only full-time employee, but it received strong support from a number of leading clinical geneticists such as Rodney Harris, Martin Bobrow, and Peter Harper, some of whom served as advisors. The group carved out an influential role by focusing on policy and advocacy, drawing on the experience of members to make complex issues understandable to policymakers whilst also feeding back policy issues into the patient community. Nonetheless, the focus was on single-gene disorders and while it was suspected that there was a genetic contribution to common complex conditions, for the moment it was seen as being 'over the horizon'.[136] In time this would change, and the place of patients and their families would come to be a significant one in and around the development of genetics and genomics policy and the debates that public health genetics would seek to facilitate.

The Cambridge Connection

By the early 1990s, with all of these developments – the evolution of genetics services, the rise of genetics as a policy issue, the burgeoning science of genomics, and the significance of ethical, legal, and social questions – thoughts began to turn to the place of genetics in relation to common complex conditions and public health more broadly defined. In Britain, Cambridge slowly emerged as a focal point for this kind of thinking, particularly under the influence of Ron Zimmern. After qualifying in Medicine at Cambridge (with training at the Middlesex Hospital), Zimmern completed junior posts in London, including at the National

Hospital for Nervous Diseases, before returning to Cambridge as a Lecturer in Medicine in 1976. Rather than applying for consultant positions in his preferred field of Neurology, however, he changed his mind about medicine and undertook a Law degree instead. A few years later he decided to train in public health. Zimmern's 'idiosyncratic' career allowed him to understand the value of working across disciplinary boundaries and developing new ideas. He was also unusual in understanding public health to be fundamentally grounded in medicine. As he describes, 'most people in public health, are interested in sort of inequalities and communities and things, I was actually interested in acute medicine .... I wasn't a conventional public health physician in that sense'. From this perspective, if clinical medicine was going to be increasingly influenced by genetics, then the same should be true of public health.

Despite Cambridge's high academic profile, the city and the surrounding East Anglian region seemed to be behind the curve in terms of the roll-out of genetics services during the 1980s. Peter Harper described the provision as 'extremely backward'.[137] This was somewhat surprising as one of local public health officers, Spencer Hagard, had written his PhD thesis on prenatal diagnosis of Downs syndrome, with Malcolm Ferguson-Smith as his supervisor and Cedric Carter as his external examiner.[138] In comparison with the leading genetic centres, there was only a relatively small health service operation. However, this did include genetic counselling clinics run by Clare Davison, who had published the first book on genetic counselling with Alan Stevenson and arrived from the MRC unit in Oxford in 1974 to help establish the regional genetics service.[139] Malcolm Ferguson-Smith's appointment in 1987 to the Chair of Pathology at the University of Cambridge also included a remit to develop medical genetic services as Director of the East Anglia Regional Genetics Service. Bruce Ponder was also appointed as Reader in Human Cancer Genetics in 1987. Yet Cambridge did not appoint its first Professor of Medical Genetics until 1995, when Martin Bobrow arrived from Guys and St Thomas's in London.[140]

Much of the innovation in Cambridge followed the appointment of Keith Peters as Regius Professor of Medicine in 1987. Having been Professor of Medicine and a Consultant Physician at the Hammersmith Hospital, Peters found Cambridge a 'rather sleepy' place. He 'lit a bonfire under it', explicitly seeking to develop a closer relationship between research and clinical services.[141] Tim Cox was appointed as Professor of Medicine in 1989, having also been a Consultant Physician – in Haematology – at the Hammersmith. Cox describes himself as 'one of Keith's boys – of which there were quite a few'.[142] The University's department of public health, a Medical Research Council biostatistics unit, and the local and regional public health departments were all brought together in one building. Peters saw this as creating a 'critical mass of people' and in time a 'powerful atmosphere' in Cambridge, which helped to lay the foundations for future developments in public health genetics.[143] Peters had got to know Zimmern when they worked together at Addenbrookes Hospital – which Peters felt was a 'very pale shadow of what a major teaching hospital should be'.[144]

Zimmern became Director of Public Health for Cambridge and Huntingdon Health Authority in 1991. Hilary Burton, who had been a public health trainee at the same time as Zimmern took up a role as a consultant in public health with responsibility for screening and children's services. She reflects that 'there were probably five, six, seven [public health] consultants in Cambridge, which is an absolutely massive luxury now. We always worked together very closely'.[145] Zimmern was keen to promote planning and the integration of research, teaching and services. A particularly formative experience was the 1995 case of 'Child B', later disclosed to be the 10-year-old leukaemia patient Jaymee Brown, which highlighted difficult issues around the cost of experimental treatment and 'rationing' in the NHS. Zimmern drew on his legal training in recommending that the Addenbrookes staff were the most appropriate to manage her palliative care even though an expensive experimental procedure with a limited chance of success was available elsewhere.[146] A central ethical point was that the needs of individuals should be weighed carefully against the need to spend resources for the benefit of the whole population.

It was around this time that Zimmern began to develop an interest in genetics. He facilitated discussions amongst the clinical community in Cambridge, at time when the developing work of Ponder, Ferguson-Smith and others was, from a small base, beginning to give genetics a new impetus – one which might begin to match that of established centres in Oxford and London.[147] Zimmern was supportive of Ponder's efforts to develop cancer genetics in Cambridge and appreciated the epidemiological and preventive dimensions of his work at a time when the association of high penetrance gene mutations with cancer were beginning to be recognised. The North East region had recently created a register for familial adenomatous polyposis under the leadership of John Burn. Anticipating its increasing relevance to public health, Zimmern encouraged one of his public health trainees, Paul Pharoah, to work with Ponder on the population significance of the BRCA1 and BRCA2 genes, which had recently been identified.[148] Pharoah is now Professor of Cancer Epidemiology at the University of Cambridge.[149]

Alison Stewart, who later became a colleague of Zimmern's, suggests that Martin Bobrow was 'responsible for Ron's conversion on the road to Damascus'.[150] Zimmern and Bobrow certainly met at the suggestion of Peters, and their conversations helped to shape Zimmern's thinking. As Bobrow recalls 'he told me that he was going to set up something called 'public health genetics', and I said, 'Oh yes, well what's that?''[151] Zimmern's intention was to try and begin a different kind of conversation, and to get involved in debates about research, the development of genetic services and genetics policy from the perspective of a non-geneticist and a public health physician. As he describes it, 'the clinicians are in their box, thinking about clinical things, and the public health people are in their box thinking about standard public health things … I really don't think that there was anybody thinking about it in this holistic way'.[152] Zimmern had been seconded into the Department of Health for two days a week when the *Working for*

Patients reforms were being developed in the late 1980s – as part of a working group on NHS Trusts, one of series which considered issues around the commissioning of health services. This gave him a useful insight into the policymaking process at a national level. Bobrow was also alert to the value of having a public health consultant advocating for more investment in genetics: 'as soon as you say, 'I am a geneticist and I think you should be doing more genetics then they discount it by half, whereas if someone says, 'I am not a geneticist, but I think you should be doing more genetics' it somehow counts for more'.[153] The confluence of key individuals and ideas in Cambridge during the 1990s was critical to the development of this new approach, which would in time have a tangible impact on the development of genetics policy in Britain.

Notes

1 C.N. Mikail, *Public Health Genomics: The Essentials* (Jossey-Bass, 2008).
2 A.H. Sturtevant, *A History of Genetics* (Cold Spring Harbor Laboratory Press, 2001). C. Yapijakis, 'Ancestral Concepts of Human Genetics and Molecular Medicine in Epicurean Philosophy' in H.I. Petermann, P.S. Harper, and S. Doetz (eds.), *History of Human Genetics* (Springer, 2017).
3 S. Mukherjee, *The Gene: An Intimate History* (Vintage, 2017) p. 25.
4 P. Harper, *The Evolution of Medical Genetics: A British Perspective* (CRC Press, 2020). P. Lanthony, *The History of Color Blindness* (Wayenborgh, 2013).
5 Harper, *The Evolution* pp. 8–9.
6 A.E. Emery, 'Joseph Adams (1756–1818)', *Journal of Medical Genetics*, Vol. 26, No. 2, 1989. p. 116. A.G. Motulsky, 'Joseph Adams (1756–1818): A Forgotten Founder of Medical Genetics', *AMA Archives of Internal Medicine*, Vol. 104, No. 3, 1959. pp. 490–96.
7 J. Slack, *Genes: A Very Short Introduction* (OUP, 2014).
8 For further details, see for example, W.S. Klug et al., *Concepts of Genetics* (Pearson, 2012).
9 Mukherjee, *The Gene* pp. 59–60.
10 R. Olby, 'William Bateson's Introduction of Mendelism to England: A Reassessment', *British Journal for the History of Science*, Vol. 20, No. 4, 1987, pp. 399–420.
11 W. Bateson and B. Bateson, *William Bateson F.R.S Naturalist: His Essays and Addresses, Together with a Short Account of His Life* (Cambridge University Press, 1928), p. 456.
12 T. M. Porter, *Karl Pearson: The Scientific Life in a Statistical Age* (Princeton University Press, 2006).
13 Mukherjee, p. 69.
14 W. Johanssen, 'The Genotype Conception of Heredity', *International Journal of Epidemiology,* Vol. 43, No. 4, 2014, pp. 989–1000.
15 J. Huxley, *Evolution: The Modern Synthesis* (Allen & Unwin, 1942).
16 Harper, *The Evolution*.
17 W. Leeming, 'Ideas About Heredity, Genetics, and 'Medical Genetics' in Britain, 1900–1982', *Studies in History and Philosophy of Biological and Biomedical Sciences*, Vol. 36, 2005, pp. 538-558.
18 Harper, *The Evolution*.
19 E.W. Straus and A. Straus, *Medical Marvels: The 100 Greatest Advances in Medicine* (Prometheus Books, 2006).
20 Slack, *Genes*.
21 Ibid.
22 Ibid.

23 Ibid.
24 I. Shine, *Thomas Hunt Morgan: Pioneer of Genetics* (The University Press of Kentucky, 1976).
25 J.D. Watson and F.H.C. Crick, 'Molecular Structure of Nucleic Acids: A Structure for Deoxyribose Nucleic Acid', *Nature*, Vol. 171, No. 4356, 1953. M.H.F. Wilkins, A.R. Stokes, and H.R. Wilson, 'Molecular Structure of Nucleic Acids: Molecular Structure of Deoxypentose Nucleic Acids', *Nature*, Vol. 171, No. 4356, 1953. R.E. Franklin and R.G. Gosling, 'Molecular Configuration in Sodium Thymonucleate', *Nature*, Vol. 171, No. 4356, 1953.
26 B. Maddox, *Rosalind Franklin: The Dark Lady of DNA* (HarperCollins, 2002).
27 Harper, *The Evolution*.
28 Mikail, *Public Health Genomics*.
29 Harper, *The Evolution*.
30 Ibid.
31 L.A. Reynolds and E.M. Tansey (eds.), *Clinical Genetics in Britain: Origins and Development. Wellcome Witnesses to Twentieth Century Medicine*, Vol. 39 (Wellcome Trust Centre for the History of Medicine at UCL, 2010).
32 Harper, *The Evolution*.
33 Ibid,p. 48.
34 C.A. Clarke, *Human Genetics and Medicine* (Arnold, 1987).
35 Reynolds and Tansey, *Clinical Genetics*, p. 38.
36 K. Dronamraju and C. Francomano (eds.), *Victor McKusick and the History of Medical Genetics* (Springer, 2012).
37 Interview with Professor Sir Peter Harper, November 2020.
38 Interview with Professor Martin Bobrow, November 2020.
39 Ibid.
40 Interview with Professor Peter Farndon, May 2021.
41 Reynolds and Tansey, *Clinical Genetics,* p. 20.
42 R. Bivins, 'Coming 'Home' to (Post) Colonial Medicine: Treating Tropical Bodies in Post-War Britain, *Social History of Medicine*, Vol. 26, No. 1, 2013, pp. 1–20. R. Bivins, *Contagious Communities: Medicine, Migration, and the NHS in Post War Britain* (Oxford University Press, 2015).
43 Reynolds and Tansey *Clinical Genetics,* p. 19.
44 S.M. Lindee, 'Genetic Disease in the 1960s: A Structural Revolution', *American Journal of Medical Genetics (Seminars in Medical Genetics)* Vol. 115, 2002. p. 75. S.M. Lindee, *Moments of Truth in Genetic Medicine* (Johns Hopkins University Press, 2005) p. 1.
45 Harper, *The Evolution*.
46 Reynolds and Tansey, *Clinical Genetics,* p. 24.
47 Harper, *The Evolution*, pp. 62–63.
48 Bivins, *Contagious Communities,* p. 322.
49 Ibid, p. 329.
50 Ibid, p. 343.
51 G. Redhead, "A British Problem Affecting British People': Sickle Cell Anaemia, Medical Activism and Race in the National Health Service, 1975–1993', *Twentieth Century British History*, Vol. 32, No. 2, 2021, pp. 189–211.
52 D.A. Christie and E.M. Tansey (eds.), *Genetic Testing. Wellcome Witnesses to Twentieth Century Medicine*, Vol. 17. 2003 (Wellcome Trust Centre for the History of Medicine). p. 16.
53 Christie and Tansey, *Genetic Testing*, p. 7.
54 Reynolds and Tansey, *Clinical Genetics*.
55 Harper, *The Evolution*.
56 M. Lambert, P. Begley, and S. Sheard (eds.), *Mersey Regional Health Authority, 1974–1994* (University of Liverpool, 2020).
57 Leeming, 'Ideas About Heredity'.

58 P.A. Coventry and J.V. Pickstone, 'From What and Why did Genetics Emerge as a Medical Specialism in the 1970s in the UK? A Case-History of Research, Policy and Services in the Manchester Region of the NHS', *Social Science and Medicine*, Vol. 49, 1999, p. 1227.

59 Interview with Professor Peter Farndon, May 2021.

60 Harper, *The Evolution*.

61 Interview with Professor Sir Peter Harper, November 2020.

62 Royal College of Physicians, *Prenatal Diagnosis and Genetic Screening: Community and Service Implications* (London, 1989).

63 Interview with Professor Sir Peter Harper, November 2020.

64 Interview with Professor Sir John Burn, December 2020.

65 Harper, *The Evolution*.

66 Leeming, 'Ideas About Heredity', p. 550.

67 Interview with Dr Rosalind Skinner, July 2021.

68 Ibid.

69 Interview with Professor Sir Peter Harper, November 2020.

70 Harper, *The Evolution*, p. 283. Interview with Dr Ron Zimmern, February 2021.

71 B. Jordan, *Travelling Around the Human Genome: An In Situ Investigation* (John Libbey Eurotext, 1993).

72 J. Archibald, *Genomics: A Very Short Introduction* (Oxford University Press, 2018) p. 18.

73 Interview with Dr David Bentley, Cold Spring Harbor Laboratory Oral History Collection.

74 Interview with Professor Sir Keith Peters, November 2020.

75 Ibid.

76 Ibid.

77 J. November, 'More than Moore's Mores: Computers, Genomics, and the Embrace of Innovation', *Journal of the History of Biology*, Vol. 51, No. 4, 2018, pp. 807–840.

78 Interview with Professor Martin Bobrow, November 2020.

79 Interview with Professor Sir John Burn, December 2020.

80 Interview with Dr Ron Zimmern, February 2021.

81 Ibid.

82 A. Mold, P. Clark, G. Millward, and D. Payling, *Placing the Public in Public Health in Post-War Britain, 1948–2012* (Palgrave, 2019).

83 C. Hamlin, 'Public Health' in M. Jackson (ed.), *The Oxford Handbook of The History of Medicine* (Oxford University Press, 2011) p. 411.

84 V. Berridge, *Public Health: A Very Short Introduction* (Oxford University Press, 2016).

85 R.A. Lewis, *Edwin Chadwick and the Public Health Movement 1832–1854* (Longmans, 1952). C. Hamlin, *Public Health and Social Justice in the Age of Chadwick: Britain, 1800–1854* (Cambridge University Press, 2008).

86 S. Sheard and H. Power (eds.), *Body and City: Histories of Urban Public Health* (Ashgate, 2000). J. Stewart (ed.), *Pioneers in Public Health: Lessons From History* (Routledge, 2017).

87 S. Sheard and L. Donaldson, *The Nation's Doctor: The Role of the Chief Medical Officer 1855–1998* (Radcliffe, 2005).

88 S. Hempel, *The Medical Detective: John Snow and the Mystery of Cholera* (Granta, 2006).

89 A.R. Omran, 'The Epidemiologic Transition: A Theory of the Epidemiology of Population Change', *Milbank Memorial Fund Quarterly*, Vol. 49, No. 4, Pt. 1, pp. 509–38.

90 J. Lewis, 'Health and Health Care in the Progressive Era', in R. Cooter and J. Pickstone (eds.), *Medicine in the Twentieth Century* (Harwood, 2000), pp. 81–95.

91 V. Berridge, *Marketing Health: Smoking and the Discourse of Public Health in Britain, 1945–2000* (Oxford University Press, 2007). S. Lock, L.A. Reynolds, and E.M. Tansey (eds.), *Ashes to Ashes: History of Smoking and Health* (Rodpoi, 1998). Sheard and LDonaldson, *The Nation's Doctor*.

92 *Inequalities in Health: The Black Report*, in Peter Townsend, Nick Davidson, and M. Whitehead (eds.), *The Health Divide* (Penguin, 1988); R. Wilkinson, *Unhealthy Societies: The Afflictions of Inequalities* (Routledge, 1996).

93 *Public Health in England: Report of the Committee of Inquiry into the Future Development of the Public Health Function.* Cm. 289 (London: The Stationery Office, 1988).

94 Mikail, *Public Health Genomics,* p. 5.

95 Hamlin, 'Public Health', p. 411.

96 L.I. Woolf and J. Adams, 'The Early History of PKU', *International Journal of Neonatal Screening,* Vol. 6, No. 3, 2020, p. 59.

97 Harper, *The Evolution.*

98 Lindee, *Moments of Truth.*

99 Interview with Dr Muin Khoury, November 2020.

100 Leeming, 'Ideas About Heredity', p. 550.

101 Interview with Professor Martin Bobrow, November 2020.

102 Lindee, *Moments of Truth,* p. 29.

103 Interview with Professor Sir John Burn, December 2020.

104 Harper, *The Evolution.*

105 N.E. Morton and C.S. Chung, *Genetic Epidemiology* (Academic Press, 1978).

106 Interview with Dr Muin Khoury, November 2020. https://blogs.cdc.gov/genomics/2013/09/12/genetic-epidemiology/

107 M.J. Khoury, T.H. Beaty, and B.H. Cohen, *Fundamentals of Genetic Epidemiology* (Oxford University Press, 1993) p. 7.

108 P. Duggal, C. Ladd-Acosta, D. Ray, and T.H. Beaty, 'The Evolving Field of Genetic Epidemiology: From Familial Aggregation to Genomic Sequencing', *American Journal of Epidemiology,* Vol. 188, No. 12, 2019, pp. 2069–77.

109 G. Rivett, *From Cradle to Grave: Fifty Years of the NHS* (King's Fund, 1998).

110 S. Shorvon, 'How the Coming of the NHS Change British Neurology', *Brain,* Vol. 141, No. 5, 2018, pp. 1570–1575.

111 R. Stevens, *Medical Practice in Modern England: The Impact of Specialization and State Medicine* (Transaction, 2009).

112 Sheard and Donaldson, *The Nation's Doctor.*

113 Interview with Professor Sir John Burn, December 2020.

114 Harper, *The Evolution,* p. 132.

115 Standing Medical Advisory Committee, *Human Genetics* (London: HMSO, 1972). Peter Harper Interview with Professor Marcus Pembrey, September 2006.

116 P. Harper, 'Paul Polani and the Development of Medical Genetics', *Human Genetics,* Vol. 120, No. 5, 2007, pp. 723–731. Reynolds and Tansey, *Clinical Genetics,* p. 57.

117 Reynolds and Tansey, *Clinical Genetics,* p. 57.

118 https://secure.eshg.org/141.0.html

119 J.S. Fitzsimmons, M. Baraitser, B.C.C. Davison, M.A. Ferguson-Smith, N.C. Navin, and M.E. Pembrey. *The Provision of Regional Genetic Services in the United Kingdom. Report of the Clinical Genetics Society Working Party on Regional Genetic Services, Supplements to the Bulletin of the Eugenic Society,* 4, 1982.

120 N.P. Halliday, 'Developments in the Health Service: The Role of the Department of Health', *Paraplegia,* Vol. 31, No. 4, 1993, pp. 199–202.

121 Peter Harper Interview with Professor Rodney Harris, March 2006.

122 E.M. Jones and E.M. Tansey (eds.), *Clinical Molecular Genetics in the UK c.1975–c.2000.* Wellcome Witnesses to Contemporary Medicine, Vol. 39. 2014. London: Queen Mary, University of London, p. 75.

123 Peter Harper Interview with Professor Rodney Harris, March 2006.

124 Jones and Tansey, *Clinical Molecular Genetics,* p. 71.

125 Ibid., p. 76.

126 Interview with Professor Marcus Pembrey, April 2021.

127 Ibid.

128 M. Dunn and T. Hope, *Medical Ethics: A Very Short Introduction* (Oxford University Press, 2018).

129 P.A. Lombardo (ed.), *A Century of Eugenics in America: From the Indiana Experiment to the Human Genome Era* (Indiana University Press, 2011).
130 'Coming Home', p. 12.
131 Harper, *The Evolution*, p. 66.
132 Interview with Professor Sir Peter Harper, November 2020.
133 Interview with Baroness O'Neill, January 2021.
134 S. Franklin, 'Developmental Landmarks and the Warnock Report: A Sociological Account of Biological Translation', *Comparative Studies in Society and History*, Vol. 61, No. 4, 2019.
135 Interview with Alastair Kent, November 2020.
136 Ibid.
137 Harper, *The Evolution*, p. 66.
138 Peter Harper Interview with Professor Malcom Ferguson-Smith, December 2003.
139 Reynolds and Tansey, *Clinical Genetics*. A.C. Stevenson and B.C. Clare Davison, *Genetic Counselling* (Heinemann Medical, 1970).
140 Peter Harper interview with Professor Martin Bobrow, November 2004.
141 Reynolds and Tansey, *Clinical Genetics*, p. 56.
142 Interview with Professor Tim Cox, June 2021.
143 Interview with Professor Sir Keith Peters, November 2020.
144 Ibid.
145 Interview with Dr Hilary Burton, February 2021.
146 C. Ham, 'Tragic Choices in Health Care: Lessons from the Child B Case', *British Medical Journal*, Vol. 319, No. 7219, 1999, pp. 1258–61.
147 Interview with Professor Tim Cox, June 2021.
148 Interview with Professor Paul Pharoah, May 2021.
149 Interview with Professor Sir Bruce Ponder, December 2020.
150 Interview with Alison Stewart, January 2021.
151 Interview with Professor Martin Bobrow, November 2020.
152 Interview with Dr Ron Zimmern, November 2020.
153 Interview with Professor Martin Bobrow, November 2020.

Bibliography

Archibald, J., *Genomics: A Very Short Introduction* (Oxford University Press, 2018).
Bateson, W. and Bateson, B., *William Bateson F.R.S Naturalist: His Essays and Addresses, Together with a Short Account of his Life* (Cambridge University Press, 1928).
Berridge, V., *Marketing Health: Smoking and the Discourse of Public Health in Britain, 1945–2000* (Oxford University Press, 2007).
Bivins, R., 'Coming 'Home' to (Post) Colonial Medicine: Treating Tropical Bodies in Post-War Britain', *Social History of Medicine*, Vol. 26, No. 1, 2013.
Bivins, R., *Contagious Communities: Medicine, Migration, and the NHS in Post War Britain* (Oxford University Press, 2015).
Christie, D.A. and Tansey, E.M. (eds.), *Genetic Testing. Wellcome Witnesses to Twentieth Century Medicine*, Vol. 17 (London: Wellcome Trust Centre for the History of Medicine at UCL, 2003)
Clarke, C.A., *Human Genetics and Medicine* (Arnold, 1987).
Cooter, R. and Pickstone, J. (eds.), *Medicine in the Twentieth Century* (Harwood, 2000).
Coventry, P.A. and Pickstone, J.V., 'From What and Why did Genetics Emerge as a Medical Specialism in the 1970s in the UK? A Case-History of Research, Policy and Services in the Manchester Region of the NHS', *Social Science and Medicine*, Vol. 49, 1999.
Dronamraju, K. and Francomano, C. (eds.), *Victor McKusick and the History of Medical Genetics* (Springer, 2012).

Duggal, P., Ladd-Acosta, C., Ray, D., and Beaty, T.H., 'The Evolving Field of Genetic Epidemiology: From Familial Aggregation to Genomic Sequencing', *American Journal of Epidemiology*, Vol. 188, No. 12, 2019.

Dunn, M. and Hope, T., *Medical Ethics: A Very Short Introduction* (Oxford University Press, 2018).

Emery, A.E., 'Joseph Adams (1756–1818)', *Journal of Medical Genetics*, Vol. 26, No. 2, 1989.

Fitzsimmons, J.S., Baraitser, M., Davison, B.C.C., Ferguson-Smith, M.A., Navin, N.C., and Pembrey. M.E., 'The Provision of Regional Genetic Services in the United Kingdom. Report of the Clinical Genetics Society Working Party on Regional Genetic Services', *Supplements to the Bulletin of the Eugenic Society*, Vol. 4, 1982.

Franklin, R.E. and Gosling, R.G., 'Molecular Configuration in Sodium Thymonucleate', *Nature*, Vol. 171, No. 4356, 1953.

Franklin, S., 'Developmental Landmarks and the Warnock Report: A Sociological Account of Biological Translation', *Comparative Studies in Society and History*, Vol. 61, No. 4, 2019.

Halliday, N.P., 'Developments in the Health Service: The Role of the Department of Health', *Paraplegia*, Vol. 31, No. 4, 1993.

Ham, C., 'Tragic Choices in Health Care: Lessons from the Child B Case', *British Medical Journal*, Vol. 319, No. 7219, 1999.

Hamlin, C., *Public Health and Social Justice in the Age of Chadwick: Britain, 1800–1854* (Cambridge University Press, 2008).

Harper, P., 'Paul Polani and the Development of Medical Genetics', *Human Genetics*, Vol. 120, No. 5, 2007.

Harper, P., *The Evolution of Medical Genetics: A British Perspective* (CRC Press, 2020).

Hempel, S., *The Medical Detective: John Snow and the Mystery of Cholera* (Granta, 2006).

Huxley, J., *Evolution: The Modern Synthesis* (Allen & Unwin, 1942).

Inequalities in Health: The Black Report, edited by Peter Townsend and Nick Davidson and M. Whitehead, The Health Divide (Penguin, 1988).

Jackson, M. (ed.), *The Oxford Handbook of the History of Medicine* (Oxford University Press, 2011).

Johanssen, W., 'The Genotype Conception of Heredity', *International Journal of Epidemiology*, Vol. 43, No. 4, 2014.

Jones, E.M. and Tansey, E.M. (eds.), *Clinical Molecular Genetics in the UK c.1975–c.2000. Wellcome Witnesses to Contemporary Medicine*, Vol. 39 (London: Queen Mary, University of London, 2014).

Jordan, B., *Travelling Around the Human Genome: An In Situ Investigation* (John Libbey Eurotext, 1993).

Khoury, M.J., Beaty, T.H., and Cohen, B.H., *Fundamentals of Genetic Epidemiology* (Oxford University Press, 1993).

Lambert, M., Begley, P., and Sheard, S. (eds.), *Mersey Regional Health Authority, 1974–1994* (University of Liverpool, 2020).

Lanthony, P., *The History of Color Blindness* (Wayenborgh, 2013).

Leeming, W., 'Ideas About Heredity, Genetics, and 'Medical Genetics' in Britain, 1900–1982', *Studies in History and Philosophy of Biological and Biomedical Sciences*, Vol. 36, 2005.

Lewis, R.A., *Edwin Chadwick and the Public Health Movement 1832–1854* (Longmans, 1952).

Lindee, S.M., 'Genetic Disease in the 1960s: A Structural Revolution', *American Journal of Medical Genetics (Seminars in Medical Genetics)*, Vol. 115, 2002.

Lindee, S.M., *Moments of Truth in Genetic Medicine* (Johns Hopkins University Press, 2005).

Lock, S., Reynolds, L.A., and Tansey, E.M. (eds.), *Ashes to Ashes: History of Smoking and Health* (Rodpoi, 1998).

Lombardo, P.A. (ed.), *A Century of Eugenics in America: From the Indiana Experiment to the Human Genome Era* (Indiana University Press, 2011).

Maddox, B., *Rosalind Franklin: The Dark Lady of DNA* (HarperCollins, 2002).

Mikail, C.N., *Public Health Genomics: The Essentials* (Jossey-Bass, 2008).

Mold, A., Clark, P., Millward, G., and Payling, D., *Placing the Public in Public Health in Post-War Britain, 1948–2012* (Palgrave, 2019).

Morton, N.E. and Chung, C.S., *Genetic Epidemiology* (Academic Press, 1978).

Motulsky, A.G., 'Joseph Adams (1756–1818): A Forgotten Founder of Medical Genetics', *AMA Archives of Internal Medicine*, Vol. 104, No. 3, 1959.

Mukherjee, S., *The Gene: An Intimate History* (Vintage, 2017).

Olby, R., 'William Bateson's Introduction of Mendelism to England: A Reassessment', *British Journal for the History of Science*, Vol. 20, No. 4, 1987.

Omran, A.R., 'The Epidemiologic Transition: A Theory of the Epidemiology of Population Change', *Milbank Memorial Fund Quarterly*, Vol. 49, No. 4, Pt.1, 1971.

Petermann, H.I., Harper, P.S., and Doetz, S. (eds.), *History of Human Genetics* (Springer, 2017).

Porter, T.M., *Karl Pearson: The Scientific Life in a Statistical Age* (Princeton University Press, 2006).

Prenatal Diagnosis and Genetic Screening: Community and Service Implications (Royal College of Physicians, 1989).

Public Health in England: Report of the Committee of Inquiry into the Future Development of the Public Health Function. Cm. 289 (London: The Stationery Office, 1988).

Redhead, G., 'A British Problem Affecting British People': Sickle Cell Anaemia, Medical Activism and Race in the National Health Service, 1975–1993', *Twentieth Century British History*, Vol. 32, No. 2, 2021.

Reynolds, L.A. and Tansey, E.M. (eds.), *Clinical Genetics in Britain: Origins and development. Wellcome Witnesses to Twentieth Century Medicine*, Vol. 39 (London: Wellcome Trust Centre for the History of Medicine at UCL, 2010).

Rivett, G., *From Cradle to Grave: Fifty Years of the NHS* (King's Fund, 1998).

Shine, I., *Thomas Hunt Morgan: Pioneer of Genetics* (The University Press of Kentucky, 1976).

Sheard, S. and Donaldson, L., *The Nation's Doctor: The Role of the Chief Medical Officer 1855–1998* (Radcliffe, 2005).

Sheard, S. and Power, H. (eds.), *Body and City: Histories of Urban Public Health* (Ashgate, 2000).

Shorvon, S., 'How the Coming of the NHS Change British Neurology', *Brain*, Vol. 141, No. 5, 2018.

Standing Medical Advisory Committee, *Human Genetics* (London: HMSO, 1972).

Slack, J., *Genes: A Very Short Introduction* (OUP, 2014).

Stevens, R., *Medical Practice in Modern England: The Impact of Specialization and State Medicine* (Transaction, 2009).

Stevenson, A.S. and Davison, B.C.C., *Genetic Counselling* (Heinemann Medical, 1970).

Stewart, J. (ed.), *Pioneers in Public Health: Lessons From History* (Routledge, 2017).

Straus, E.W. and Straus, A., *Medical Marvels: The 100 Greatest Advances in Medicine* (Prometheus Books, 2006).

Sturtevant, A.H., *A History of Genetics* (Cold Spring Harbor Laboratory Press, 2001).

Watson, J.D. and Crick, F.H.C., 'Molecular Structure of Nucleic Acids: A Structure for Deoxyribose Nucleic Acid', *Nature*, Vol. 171, No. 4356, 1953.

Wilkins, M.H.F., Stokes, A.R., and Wilson, H.R., 'Molecular Structure of Nucleic Acids: Molecular Structure of Deoxypentose Nucleic Acids', *Nature*, Vol. 171, No. 4356, 1953.

Wilkinson, R., *Unhealthy Societies: the Afflictions of Inequalities* (Routledge, 1996).

Woolf, L.I. and Adams, J., 'The Early History of PKU', *International Journal of Neonatal Screening*, Vol. 6, No. 3, 2020.

2
FOUNDATIONS

Influential Acronyms

At a press conference in October 1988 James Watson, the first Director of the Human Genome Project (HGP) at the US National Institute for Health (NIH), surprised observers by announcing that a percentage of the project's annual budget would be devoted to research into its ethical, legal, and social implications (ELSI). The importance of thinking through such issues had been established in a number of reports, but Watson's endorsement 'effectively ratified' previous recommendations.[1] The mechanism for distributing funding was developed by the 'ELSI working group' – a sub-committee of the US National Advisory Council on Human Genome Research, which administered funding on behalf of the NIH and the Department of Energy – and 3% was allocated to research into these kinds of questions. That figure later rose to 5%. The initial research agenda outlined in 1990 identified nine key principles or aims:

- Fairness in insurance, employment, the criminal justice system, education, adoption, the military, and other areas.
- Psychological and societal responses to individual genetic information.
- Privacy and confidentiality (including ownership, control, and consent).
- Genetic counselling, including pre-natal and pre-symptomatic testing, testing in the absence of therapeutic options, and population screening versus testing.
- Issues of reproductive choice.
- Medical practice, including standards of care, training, and education of patients and the general public.
- Historical misuses of genetics, especially eugenics, and relevance to the present.

DOI: 10.1201/9781003221760-3

- Commercialization, including property and intellectual property rights, and accessibility of data and materials.
- Philosophical issues such as definitions of health and disease and questions of determinism and reductionism.[2]

This approach took account of issues that had been important throughout the somewhat 'checkered history' of genetics, particularly its more 'socially unjust' elements. As Eric Juengst, the first Director of the ELSI programme, has described, 'members did not need to turn to science fiction for direction in establishing its initial agenda: they only had to draw from their own experience in [twentieth century] human genetics to identify challenges that could be reliably anticipated from the Human Genome Project'.[3]

The ELSI initiative quickly grabbed the attention of a wide range of academics and researchers. It provided a crucial source of funding for those thinking about genetics outside of the usual clinical environments – in fields such as the humanities, social sciences, and law – and was 'jaw-droppingly unique' as a result.[4] The potential implications of the project were huge, encompassing patients, clinicians, policymakers, legislators, and educators. Such an approach, with the potential for self-criticism built in, had not been widely attempted before.[5] Much of the early funded research focussed on the quality and efficacy of genetic tests, the possibility of genetic discrimination in employment and insurance, the handling of genetic information and scientific education.[6]

A self-defined 'ELSI community' developed. According to Wylie Burke, now Professor Emeritus at the Department of Bioethics and Humanities at the University of Washington, 'these scholars may have come to genetics from a variety of different places, but the availability of funding made what might otherwise have been ideas among other ideas become something they could focus on'.[7] Over time, ELSI research became 'institutionalized' into wider US academic and policymaking structures. Indeed, observers suggest that it has had a 're-markable' and 'unexpected' multidisciplinary impact on the conduct of genomic research itself and become a model that has been picked up by other countries.[8] In Canada, a similar agenda was supported in the 1990s by MELSI funding – Medical, Ethical, Legal, and Social Implications – and taken further from 2000 by Genome Canada with its GE3LS research on Genetics, Ethical, Economic, Environmental, Legal, and Social issues. A number of initiatives around the ethical, legal and social aspects were also developed in Europe.

In Britain there was a more established 'nexus' of academics and clinicians thinking about these kinds of issues. But they were few in number and the United States was seen as having a 'deeper bench'.[9] The leading thinker about the ethical implications of the 'new genetics' in Britain was Professor Martin Richards of the University of Cambridge, who published the influential book *The Troubled Helix* with Theresa Marteau in 1996.[10] The Science and Technology Committee of the House Of Commons argued in its 1995 report *Human Genetics: The Science and its Consequences* that there was 'an urgent need for more research into the Ethical,

Legal and Social Implications of genetics'.[11] Funding on a small scale had been provided by the Medical Research Council (MRC) and the Economic and Social Research Council (ESRC). The MRC subsequently considered introducing its own MELSI programme, before the ESRC funded Genomics Network began in 2002.[12]

These research opportunities across institutional and disciplinary boundaries provided important foundations for the subsequent development of public health genetics. ELSI principles were built on and taken further. Having studied for a PhD in genetics in Seattle – the University of Washington was the 'world centre of yeast genetics' – Burke undertook an MD and moved into clinical medicine and primary care. As an academic from 1988 she was then able to bring the two elements together through research. As she describes:

> The question for me was always 'where does genetic medicine fit in in routine clinical care?' I understood where it fit in in the care of individuals with unusual genetic diseases, but where does it fit in in routine clinical care?[13]

Burke worked on an ELSI funded project that considered the implications for patients of information produced by genome sequencing, and became a member of the 'ELSI community' after being part of research groups that examined ethical issues around the genetics of breast cancer and hemochromatosis.[14] Studies included patient perceptions of genetic testing, the place of counselling and screening evaluation.[15] Over the course of the next few years, she met Muin Khoury, Ron Zimmern and others and their shared interests became clear. The ways in which public health genomics subsequently evolved are considered in more detail in Chapters 3 and 4, but the early intention of those pioneers was that the field would seek to address many of the same issues as the ELSI programme, aiming for better integration between the science of genetics and its clinical implications and ensuring the benefits were felt in terms of population health.

The work of the Human Genome Organization (HUGO), created just as the Human Genome Project was getting off the ground, was also influential. Established in 1988 under the influence of Sydney Brenner and colleagues, HUGO began as a coordinating body which oversaw international efforts around gene mapping and genome sequencing, encouraging collaboration. Victor McKusick served as the first President. HUGO was conceived as a 'UN for the human genome', and one of its initial aims was to 'encourage public debate and provide information and advice on the scientific, ethical, social, legal, and commercial implications of human genome projects'.[16] An ethics committee was one of six HUGO advisory groups which were established and met for the first time in 1992. It was explicitly international in make-up and outlook and included individuals from a range of disciplines. Having initially had something of a 'freewheeling atmosphere', the committee began to produce a series of statements on issues that it considered important:

1996	Principled Conduct of Genetics Research
1998	DNA Sampling: Control and Access
1999	Cloning
2000	Benefit Sharing
2001	Gene Therapy Research
2002	Human Genomic Databases
2004	Stem Cells
2007	Pharmacogenomics (PGx): Solidarity, Equity, and Governance
2013	Imagined Futures: Capturing the Benefits of Genome Sequencing for Society.[17]

Those involved now see many of the statements as being remarkably prescient – a 'precursor' to or 'harbinger' of future debates.[18]

Running through them was a leitmotiv about the human genome as a shared public resource – part of the 'common heritage of humanity' – the benefits and risks of which should be shared equally.[19] The first statement on the 'Principled Conduct of Genetics Research' published in 1996 identified ten key principles: competence, communication, consultation, consent, choices, collaboration, conflict of interest, compensation, and continual review.[20] The 2000 statement on 'Benefit Sharing' delineated ideas of community and justice and went as far as to recommend that 'profit-making entities dedicate a percentage (e.g. 1–3%) of their annual net profit to healthcare infrastructure and/or to humanitarian efforts'.[21] The ethics committee was described by Gert-Jan van Ommen, the President of HUGO between 1998 and 1999, as the 'jewel in the crown'.[22] From 1996 to 2004 its Chair was Bartha Knoppers – then Professor of Law at the University of Montreal, now Canada Research Chair in Law and Medicine at McGill University. Knoppers had first studied comparative literature before moving on to medical law and came to genetics through an interest in reproductive technologies. In 1985, she was invited by Charles Scriver and Claude Laberge – 'the top Canadian paediatric geneticists' – to help develop the policy side of newborn screening programmes in Quebec.[23] Laberge had helped pioneer neonatal screening in the Province during the 1970s. To Knoppers, this work demonstrated the importance of thinking collectively across disciplinary boundaries about population health:

> I went to scientific meetings ... I swear I didn't understand a thing. I knew nothing about what a chromosome was. I knew Mendel's peas from Grade eight, but I learned from the scientists. I tell my students today ... 'go to the scientific meetings and sit there and listen, because if you don't understand the science then you are just inventing issues and you are no help to anyone'.[24]

By the late 1990s, the thinking being done by Knoppers and the HUGO ethics committee was complementary to that of Burke, Khoury, Zimmern and others, and fruitful interactions followed.

This kind of approach was also picked up by UNESCO – the United Nations Educational, Scientific and Cultural Organization tasked with promoting international cooperation. The pace of developments in genetic science had prompted moves to define and protect the legal status of the human genome and in 1993 UNESCO established an International Bioethics Committee, chaired by Noelle Lenoire. Lenoire was a member of the French Constitutional Court and had recently led a review of French bioethics law. The committee was made up of fifty 'independent experts in anthropology, biology, genetics, law, medicine and philosophy, chosen … to reflect the geographical and cultural diversity of UNESCO'.[25] Drafting of a declaration was undertaken by a Legal Commission, chaired by the Uruguayan jurist Héctor Gros Espiell, which convened meetings with academic experts from 1995. Knoppers was a member of both the committee and the commission, which went through a series of nine drafts over two years in developing a declaration. Government representatives from UNESCO member states then negotiated the final language in 1997. At the time, the United States was not a member of UNESCO, but it sent Eric Meslin, the Research Director of the ELSI programme, as an observer. Meslin had been a 'consulting clinical bioethicist' and an academic in Toronto before joining ELSI in 1996. In 1998, he was appointed by President Clinton as the Executive Director of the National Bioethics Advisory Commission, which provided advice about issues such as cloning and stem cell research.[26] Meslin is now President of the Council of Canadian Academies, and from the mid-2000s he became a frequent collaborator with Zimmern and the other members of the public health genomics community.

In November 1997, the Declaration on the Human Genome and Human Rights was adopted by UNESCO and then endorsed by the UN General Assembly in early 1998. The declaration provided the first ethical framework in the field of genetics at an international level. A degree of stability was sought in a fast-moving field, but the need for a broad measure of consensus and the fact that individual countries would still be responsible for genetics in their own jurisdictions meant that the focus was on principles rather than the regulation of specific practices.[27] The birth of Dolly the sheep via cloning in 1996 had prompted complex debates about the best way to address the possibility of human cloning, and a line was inserted into the declaration to the effect that this would 'not be permitted'.[28] Cloning was perceived to run counter to the declaration's central principle of maintaining human dignity. The genome was seen as 'the heritage of humanity' which 'underlies the fundamental unity of all members of the human family'.[29] There was also plenty of familiar language around the importance of consent, confidentiality, and discrimination. Perhaps most strikingly, however, emphasis was placed on the essential individual nature of the human genome, which was not fixed but subject to varying environmental factors: 'The human genome, which by its nature evolves, is subject to mutations. It contains potentialities that are expressed differently according to each individual's natural and social environment, including the individual's state of health, living conditions, nutrition and education'.[30] This was the context in which genetic medicine and genetics policy, including in Britain, were being developed during the 1990s. Ideas and perspectives that would become central to public health genomics were increasingly well established.

A Policy Foothold

Having made most of the running on the development of genetics services and accompanying policy themselves, from the mid-1990s clinical geneticists in Britain began to be able to draw on a more supportive policy infrastructure. More interest in and responsibility for genetics began to emanate from the centre. An important place in the landscape was taken up early on by the Nuffield Council on Bioethics, established in 1991. While many countries had established a national ethics committee – in France the Comité Consultatif National d'Ethique had been formed in 1983 – such a move in Britain had long been resisted by the government. It was feared that ethical questions that were not currently the responsibility of Ministers would become politicised.[31] According to Peter Harper, officials 'washed their hands of it'.[32]

The first Chair of the Nuffield Council was Sir Patrick Nairne, former Permanent Secretary at the Department of Health and Social Security between 1975 and 1981. By the late 1980s, during intense debates about the future of the NHS, Nairne emphasised the need to further develop public trust. As he described, 'no public service thinks less about the public than the NHS'.[33] His call dovetailed with concerns about the rapid development of genetics. More genetic tests meant more patient engagement and more ethical questions. There were particular concerns about public understanding of genetics and the nature of political debates around issues such as embryo research and abortion, which were seen by Stephen Lock the Editor of the *British Medical Journal,* for example, to lack the informed underpinning that might be provided by a national ethics committee.[34] There was also a perception that the scale of future ethical questions was likely to be such that practising geneticists could no longer be left to work things out for themselves without recourse to a wider framework of principles or being the subject of more scrutiny. With limits to political will, however, bodies like the Nuffield Council were able to fill the gap, presenting themselves as intermediaries and bringing interest groups together to think through difficult questions.[35]

When it became clear that a national ethics committee would not be formed, influential voices like Nairne and David Weatherall looked to the Nuffield Foundation – an independent charity founded in 1943 that funded scientific and medical research – for support. A series of initiatives was put in motion, including a consultation with a wide range of interest groups and 60 leading academics. Baroness O'Neill recalls receiving a letter from Nairne:

> I answered his letter … and wrote him a couple of pages on why I thought these were important questions, and I suppose, because I had taken the trouble to do that, I found myself a year later being asked to be a member of the new council.[36]

At the time O'Neill was Professor of Philosophy at the University of Essex and would go on to build on her reputation as a leading thinker on issues around consent, trust, and accountability. Alongside Weatherall and other representatives of ethics, philosophy, science, and medicine, she became one of the founding members and later became Chair in 1996. The Nuffield Council was funded by the Nuffield Foundation for the first three years, and then by a combination of the Nuffield Foundation, the Medical Research Council, and the Wellcome Trust. The aim was to identify ethical issues in research likely to be of public interest and provide relevant guidance. As the most widespread area of genetics provision, screening was the first issue selected and a working group of clinicians and academics, which included Peter Harper, was brought together to provide advice.

The Nuffield Council's first report *Genetic Screening – Ethical Issues* was published in 1993.[37] It added to a growing body of literature at a time when the potential misuse of genetic information was of increasing popular concern.[38] The main recommendations centred on ensuring informed consent, introducing guidelines for the retention and disclosure of genetic information, caution around employment and insurance, and improved public understanding of genetics. The most ambitious proposal was for the Department of Health to establish 'a central coordinating body to review genetic screening programmes and monitor their implementation and outcome'.[39] Many of the Nuffield Council's subsequent reports also touched on issues relevant to genetics, including human tissue and stem cell therapy. The historian Duncan Wilson has cast some doubt on the Council's ability to influence policymaking, suggesting that while its independence had advantages, it also meant that there was no real need for the government to act on its recommendations.[40] For example, a Human Genetics Advisory Committee was eventually established by the government in 1995, but only after the influential report by the House of Commons Science and Technology Committee had also made the case for having such a body. Nonetheless, Peter Harper and other observers have maintained that the reports produced by the Nuffield Council were useful to geneticists and helped to move the agenda forward.[41] The Council has also clearly been seen as the effective British equivalent of formal national ethics committees on the international stage.

David Weatherall had also been a member of the Clothier Committee, set up by the government in 1990 to examine the ethics of gene therapy. Advances in molecular biology meant that it would be increasingly possible to isolate particular genes and modify them in order to treat or prevent genetic disorders. This would clearly have implications for a number of ongoing ethical debates, including around scientific research and parental choice. Chaired by the lawyer Sir Cecil Clothier – who had led a number of parliamentary investigations and served as Health Service Commissioner for England, Scotland, and Wales between 1978 and 1985 – the committee set out in its 1992 report the ways in which a genetic dimension heightened the ethical issues inherent to any new medical intervention but noted that the issues themselves were not necessarily new. It was argued that gene therapy 'should be directed to the alleviation of genetic disease in individual patients' but not 'used to change or enhance normal human traits'.[42] Modification of body cells was permissible, but the modification of germline cells was not. Here the committee picked up on wider public concerns about the implications of scientific progress, or 'irrational fears which derive from misunderstandings of biology, and are compounded by the effects of popular creations of fiction, such as Frankenstein's monster'.[43]

The Nuffield Council recommended that a new supervisory body should be established to help local ethics committees and advise about safety and efficacy.[44] The science of gene therapy itself was still at an early stage, making the transition from scientific research to medical practice, and there was a balance to be struck between legislating and stifling important and potentially life-changing research. There were some concerns inside the Department of Health about the sensitivities of the issue, not least the potential for overlap with the remit of the recently established Human Fertilisation and Embryology Authority, though gene therapy was also understood to be less ethically fraught than some other genetic issues such as screening.[45] As the *New Scientist* observed, 'few people disagree with the principle of treating sufferers of incurable disease as long as the treatment is safe'.[46] The Gene Therapy Advisory Committee which was subsequently established in 1993 was first chaired by the leading paediatrician Professor Dame June Lloyd, who had also served on the Clothier committee. There was also continuation in the form of Martin Bobrow, who was one of the geneticists represented, and others, including the broadcaster Nick Ross who became a lay member. In practice safety concerns were often more prominent than ethical issues and Ross saw the group as 'promoters rather than restrainers' of new medical techniques.[47] Though its role has changed, the Gene Therapy Advisory Committee remains part of the regulatory landscape. Its early influence may have been most telling in terms of raising awareness of genetics inside the Department of Health and demonstrating that ethical questions would be central to the development of the field.[48]

Genetics was most keenly felt as a policy issue in terms of the provision of existing genetics services. A central government review was initiated in 1993, with the report *Population Needs and Genetic Services: An Outline* Guide produced by the Genetic Interest Group and sent out to health service managers and commissioners, alongside a joint 'Professional Letter' from the Chief Medical Officer and the Chief Nursing Officer, Kenneth Calman and Yvonne Moores, to help them understand the issues and assess provision in their area.[49] The report described the nature and impact of a number of genetic conditions and set out the likely needs of service users, using a hypothetical district of 250,000 people. For example, there would be approximately 25 patients with cystic fibrosis and 8 with muscular dystrophy in such an area. Genetic knowledge was seen to be rapidly expanding. Genetics was expected to become 'an inseparable part of health care', influencing every clinical speciality.[50]

A key question, therefore, and one that would be regularly returned to, was whether the regional service patterns that had developed on the ground would be able to meet increased demand and changing patient needs. In more specific terms, a working party of the Standing Medical Advisory Committee produced a report on services related to sickle cell disease, thalassaemia and other haemoglobino-pathies in 1993.[51] Blood diseases had long had an important place in the development of medical genetics, but until the 1980s conditions such as sickle cell disease had been 'largely invisible in Britain'. Understanding of the disease was often racialised because it disproportionately affected some ethnic minorities, and treatment and services remained under-developed as a result.[52] The key role in driving service improvements was played by community pioneers such as Elizabeth Anionwu and Milica Brozovitch. The Sickle Cell and Thalassaemia Centre opened in London in 1979 brought together screening, counselling, and treatment services. However, as with other genetic conditions, there was a need to go further.

The Standing Medical Advisory Committee report called for the much wider identification of the carriers of sickle cell disease and better provision of antenatal and neonatal screening. It recommended that specialist counselling should be fully integrated, and services should be universal in local districts in which more than 15% of the population were from ethnic minorities.[53] This work further demonstrated that there was increasing interest in genetic conditions and the provision of services inside the Department of Health. However, a later review suggested that the 1993 recommendations had not been widely implemented because they were presented to the NHS as guidelines rather than formal policies. The uncertainty of the purchaser–provider split after 1991, and the lack of extra funding to actually facilitate better services meant that little changed on the ground.[54]

While the Department of Health was just beginning to get to grips with genetics, clinicians and researchers were continuing to push the field forward. The development of Polymerase Chain Reaction (PCR) techniques made it much easier to identify and amplify genes. Single-gene disorders were increasingly well understood as a result, with around five thousand disease genes identified by the mid-1990s. Significant possibilities opened up after the discovery of the gene mutation associated with Huntingdon's disease in 1993, for example. Tests for a wider range of genetic conditions were becoming available, bringing with them a range of ethical questions. At the same time, the conversation was slowly turning towards the potential for understanding the genetic contribution to complex common conditions such as cancer, diabetes, cardiovascular disease, and Alzheimer's disease. Clothier and the Genetic Interest Group had anticipated that more would be understood about the underlying mechanisms of disease, particularly in terms of the interactions between genes and between multiple genes and external factors. Many leading geneticists had anticipated these kinds of developments. The premise was part of Weatherall's analysis in *The New Genetics and Clinical Practice* and that of Connor and Ferguson-Smith in *Essential Medical Genetics*, key texts which went through several iterations during the 1980s and 1990s.[55] Earlier detection, better prediction, and more effective treatments for common diseases were likely soon to be possible. In turn, this raised significant practical questions for the NHS. There was huge potential, but also a number of important issues that would have to be carefully thought through.

The Department of Health's engagement with these issues was demonstrated by the publication of two reports in 1995 which discussed the 'new genetics'. The first was produced by the Genetics Research Advisory Group, led by Martin Bobrow, for the NHS Central Research and Development Committee, on the development of genetics as a whole. The second was produced by another working group, led by John Bell, on the genetics of common diseases. The Canadian-born Bell had been a Rhodes Scholar and undertaken postgraduate medical training in Britain before moving in 1982 to Stanford University – the leading centre for DNA sequencing at the time. He then returned to the University of Oxford in 1987 and took the Chair of Clinical Medicine in 1992. He credits David Weatherall with getting him 'excited about molecular medicine'.[56] With support from the Wellcome Trust, Bell and colleagues established the Wellcome Centre for Human Genetics in Oxford in 1994, with the aim of more directly investigating the genetic contribution to common complex conditions. The Canadian Mark Lathrop was appointed as the first Scientific Director. This was far from the fully formed genomic science that would develop later, but it was an important antecedent. According to Bell, 'At that stage we were busy mapping genes in common diseases using affected sibling pairs' – the process of comparing relevant chromosome regions in phenotypically similar siblings – 'we hadn't yet got to the point where you could do large scale genome wide association studies, because we didn't have enough markers'.[57]

The two reports were commissioned as part of the NHS Research and Development Programme. The Programme Director was Professor Sir Michael Peckham, a leading British oncologist, who would go on to help develop important arms-length bodies such as the National Institute for Health Research and the National Institute for Clinical Excellence. During the mid-1990s, the government had outlined a policy agenda around science and technology, overseen by William Waldegrave, who served as Chancellor of the Duchy of Lancaster and Minister for Public Service and Science. This included the 1993 White Paper *Realising our Potential: A Strategy for Science, Engineering and Technology* which aimed to maintain and build on British strengths.[58] Genetics had been identified as a growth area. It offered many clinical benefits, but it could also contribute to 'wealth creation'. As Peckham wrote, 'Advances in genetics offer the prospect of considerable gain not only for the NHS but also for the British pharmaceutical and health care industries, with attendant benefits of employment, exports and growth in national income'.[59]

The crux of the first report led by Bobrow was the need to, for the moment, prioritise research on single-gene disorders. Advances in genetics were initially expected to impact testing, screening, and, potentially, new therapies for inherited diseases. The report called for consolidation of progress on conditions such as cystic fibrosis, muscular dystrophy, and Down's syndrome. For example, there was likely to be a significant and welcome shift in screening for Down's syndrome from the second to the first trimester of pregnancy – though there was an

Oxford, and Ian Lister Cheese from the Department of Health. The original group led by Bobrow had been more clinical in nature. It had included geneticists such as Rodney Harris, Nicholas Wald, and Frances Flinter, as well as a number of molecular biologists. The academic health psychologist Theresa Marteau, an expert in patient behaviour and understanding of risk, was a member of both groups.

The two reports were published concurrently in 1995, with Peckham suggesting that the second had 'been able to build on and complement the earlier study'.[66] Bobrow's perception was that the second working group had been formed – 'without initially telling me' – because Peckham had 'decided that the rather pragmatic and downbeat messages coming out of this were so far from what he wanted'. Rather than concentrate on health delivery, the Department of Health preferred 'the fantasy of the year after tomorrow'.[67] While their approach might have been more amenable to Peckham and government ministers, the problem for Bell and those who shared his perspective was that ultimately there was still a lack of data to support their vision. It was agreed that advances in genetics would impact clinical medicine, but it was not yet possible to say with any real certainty when and how this would happen As Bell describes, 'we didn't really have a story that was going to go beyond mendelian genetics'.[68] Peter Harper's view was that by the time the two reports were published the second report had 'more or less collapsed'.[69] Bell thought the impetus had 'petered out'.[70] Even so, the fact that the Department of Health was interested in these kinds of questions was significant. Policymakers were aware of the changing nature of the field. This included practical issues like the future organisation of genetics services, but also the potential for significant changes in understanding of disease and the economic opportunities that might open up as a result. This dichotomy, between advances in relation to genetic disorders and advances in relation to common complex conditions, has continued to frame important debates. Yet it wasn't until the 2010s, once genomic science and associated technologies had further developed, and the political and clinical contexts had become more propitious, that Bell's vision became more influential.

The need to prepare for the future was also the theme of a report produced by the Welsh Health Planning Forum in 1995, which saw Britain as being 'on the verge of major technological changes in health care'.[71] At a workshop which brought together 40 clinicians, academics, and NHS administrators, the Forum's Executive Director, Morton Warner, had asked the key question, 'How big is this thing going to get?'[72] It was assumed that advances in genomic science would start to feed through into health service activity within the next five to ten years. It was less clear who should take the lead on the necessary policy development. Tony Beddow, Chief Executive of West Glamorgan Health Authority, suggested that politicians were still 'relatively uninterested' and 'horribly uninformed'. David Hunter, Professor of Health Policy and Management at the University of Leeds, was actually 'glad that a diverse, fragmented society made it difficult to impose a centralised view', while Michael Jeffries, a GP and Medical Director of the Clwydian Community Care NHS Trust, predicted that relevant legislation would only arrive 10 to 15 years after best practice had been established in the field.[73] There was still some reticence therefore about the centre taking a more direct interest in genetics. A 1995 report by the Genetic Interest Group focused on the extent to which the everyday reality of the health service might impinge on the development of genomic medicine. If its promise was to be realised then significant issues around underfunding, variation in provision, and difficulties in planning would have to be addressed:

> The genetic service as it exists today in the UK is the outcome of individual initiatives and interest, rather than a consequence of a strategic planning process of any description. Typically, a specific service will have emerged out of a research project at an academic institution.[74]

In particular, the 1991 purchaser–provider split arrangements were seen to have been 'a step backwards as compared with the old regional arrangements' for commissioning, which actively 'threaten to undermine the effectiveness of the service' and 'carry with them the danger of fragmentation'.[75] One way of ensuring quality services, the report suggested, would be to introduce a national appraisal scheme. There should also be more oversight and coordination from the centre through the creation of a specific department within the NHS with responsibility for genetics and genetics services.[76]

A number of issues were coming to a head – the potential for a transformation in understanding of disease, the need to ensure the provision of high-quality genetics services, and the ethical, legal and social issues which continued to surround them. The increasing recognition by policymakers that they would have to respond was demonstrated most clearly by the publication of the report *Human Genetics: The Science and its Consequences* by the cross-party Science and Technology Committee of the House of Commons in 1995.[77] The Committee was initially drawn to genetics because of its role in overseeing the work of the MRC, which provided much of the research funding in the field. Its remit was more closely aligned with that of the government Office for Science and Technology, established in 1992, than the Department of Health.

The chair of the Committee, the Conservative MP Sir Giles Shaw, later described how, 'We believed that this problem was urgent because tales of square tomatoes and genetically induced monsters began to plague the public press'.[78] The Committee took evidence from a wide range of interest groups, including the MRC, the Department of Health, the insurance industry, religious groups, and patients' organisations, as well as a number of leading clinicians. Witnesses included Martin Bobrow, Peter Harper, Alastair Kent, Walter Bodmer, David Weatherall, Sydney Brenner, Patrick Nairne, and June Lloyd. Committee members visited genetics centres in Cambridge, Oxford, Edinburgh, and Cardiff, and undertook trips to Europe and the United States, where they met with Francis Collins, who had succeeded James Watson as Director of the National Centre for Human Genome Research. They also visited a number of US research centres and biotechnology companies. Two geneticists, David Porteous of the University of Edinburgh, and Bryan Sykes of the University of Oxford served as special advisors.

Human Genetics: The Science and its Consequences saw the field as being at a 'watershed' moment:

> Some diseases have been found to be associated with a single faulty letter in the genetic code of the embryo. But most diseases, and all the common ones which have a genetic factor, are thought to be due to the complex interaction of several genes and their changing environment.[79]

That genetic knowledge might, in time, change the face of medicine itself was well understood, but the report also sounded a note of caution:

> The scientific discoveries made increase our understanding of ourselves; of 'what make people what they are', and, some argue, will change that understanding. But while the power for good that genetics should have is acknowledged there are also frequently expressed fears that genetic knowledge could be used badly … the challenge is to ensure that, as far as possible, we gain the benefits while avoiding the abuses.[80]

In discussing the limits of genetic determination, the report drew on the words of the Reverend Dr John Polkinghorne:

> Biology has scored its first stunning, absolutely stunning quantitative success in molecular genetics … The tendency in that upbeat phase is to say that what has solved this problem will solve every problem. We should always be suspicious of intellectual passkeys that open every lock in that sort of way.[81]

As in other significant reports produced during this period, a wider framework of human rights was thought to be fundamentally important. The committee saw genetic information as being different from other medical information, and in the context of issues such as testing, screening, insurance, and employment, the emphasis was on ensuring privacy, choice, and informed consent. The importance of genetic counselling being paired with screening initiatives, particularly in the context of early diagnosis of late-onset conditions was made clear. It called for a new mechanism for evaluating genetic screening through a 'national quality assurance programme'. The committee recognised the importance of the ELSI program in the United States. Given the different funding structures, it would not be sensible to have a fixed percentage of research budgets devoted to ELSI issues in Britain, it argued, but there was an 'urgent' need for more ELSI-type research to ensure that debates about genetics were well informed. The Committee noted that 'genetic knowledge means that ethical dilemmas, rather than being infrequent special cases, may face us all the time'.[82]

In more practical terms, the committee recognised that increased spending on genetics would likely mean reductions in other health service and research areas, but the wider economic and industrial benefits were such that funding should at least be maintained at current levels, not least in relation to Britain's contribution to the Human Genome Project. The committee also recognised geneticists concerns about the impact of the purchaser–provider split:

> Many of our witnesses expressed concern that the NHS reforms would make the collection of family material and the referral of patients to research centres more difficult. This would not only hinder research but prevent patients and their families from receiving the best care possible.[83]

Although the report ultimately concluded that:

> It is too soon to determine the weight to be attached to this view. Accordingly, the effects of the NHS reforms on genetic research should be carefully monitored.[84]

Nonetheless, there was a clear desire for more central oversight and coordination and the report recommended the establishment of a new Human Genetics Commission – reflecting the 1993 recommendation from the Nuffield Council on Bioethics. The Science and Technology Committee proposed a wider regulatory remit, covering all genetic medicine and research:

We recommend that there should be a body (The Human Genetics Commission) to

- monitor the availability of genetic services in different regions.
- advise Local Research Ethics Committees on research involving genetic screening.
- approve screening programmes before they are introduced.
- disseminate best practices and keep it under constant review.
- monitor the availability of genetic diagnosis and screening and make recommendation to local purchasers, if appropriate.
- prescribe the circumstances in which particular types of screening or diagnosis, such as pre-natal diagnosis, should be provided or proscribed.[85]

The scope of the new Commission's responsibilities would be such that it would need to be established on a statutory rather than a voluntary basis.

By the mid-1990s therefore, a more supportive policy infrastructure for genetics was beginning to develop, and there were increasing political and professional calls for more coordination and regulation from the centre. Government, however, remained reluctant to act. Peter Harper, who had been visited by the Committee in Cardiff, reflected simply that the government 'didn't want to know'.[86] In its official response to the committee, the government promised to keep the situation under review but declined to establish a commission on the lines recommended.[87] The issues were thought to be covered by existing advisory bodies. Where there was a gap, an Advisory Committee on Genetic Testing (ACGT) would be introduced, with Polkinghorne becoming its first Chair. This would be able to suggest some changes, but it lacked formal regulatory powers. The government's response was very much framed in the context of the wider policy agenda around science and technology, and the need to ensure competitiveness and maintain British leadership. Further central oversight in the development of genetic services or regulation of research was not thought to be desirable.

Members of the Science and Technology Committee did not take the government's response lying down. They established another 'mini-inquiry' and produced a further report which set out their concerns again, particularly around the regulation of genetic tests and the implications for insurance, which the government had argued was essentially a matter for employers and the insurance industry.[88] Leading clinicians were also robust in their criticism, suggesting that the government had failed to understand the fundamental nature of the ongoing scientific and technological changes. At the time, Harper described the government's response as 'abject spinelessness'.[89]

Under pressure from MPs and clinicians, the government did subsequently accept that a larger independent body might have a role to play. The Minister for Science and Technology, Ian Taylor, described a process of 'co-operation and inspired agreement as a result of further deliberations'.[90] A Human Genetics Advisory Commission (HGAC) was then established in 1996. However, it lacked regulatory teeth. The Government's second official response suggested that 'it is difficult to see how such a body could avoid becoming unwieldly and bureaucratic. This would be particularly inappropriate in a field where the science is advancing rapidly, requiring flexibility of response'.[91] In explaining the reluctance to intervene more directly, Taylor argued that 'The Government would be the worst type of regulator in such a hugely complicated market'.[92]

The Advisory Committee on Genetic Testing, set up in response to the committee's first report, did pick up many of the same social and ethical issues around testing that had been highlighted in earlier reports. Alongside Polkinghorne, the Committee included established figures such as Peter Harper, Marcus Pembrey and Hilary Harris. The journalist and former Conservative MP Matthew Parris served as a lay member. The focus was often on developing relevant guidelines or codes of practice. A 1997 report on direct-to-consumer genetic tests which established professional standards and emphasised the need for adequate genetic counselling alongside testing was 'well received', though there was no statutory underpinning.[93] Harper thought that this committee did some 'really useful, practical, down to earth things' within its 'fairly limited remit'.[94]

The Human Genetics Advisory Committee (HGAC), set up in response to the second report, sought to consider genetic issues outside of the direct context of healthcare, review scientific progress and build public understanding. Dolly the Sheep cast a long shadow, and in 1998 the Committee published a joint report with the HFEA which argued for maintaining a ban on human reproductive cloning.[95] The HGAC was conceived as a 'strategic group of independent members'.[96] The first Chair was the lawyer Colin Campbell, followed by Baroness O'Neill. The broadcaster Moira Stuart served as a lay member. Insurance quickly became an area of interest. Its 1997 report *The Implications of Genetic Testing for Insurance* sought to resolve issues that had been going back and forth for some time, including between the Government and the Science and Technology Committee.[97] The HGAC's approach was to say that while insurers were entitled to an individual's genetic information in principle, just as they were entitled to other medical information, this should be dependent on there being a clear scientific case for that information to be used in the setting of insurance premiums. The HGAC proposed a two-year moratorium on the use of genetic test results by insurance companies, while an independent supervisory body and a transparent appeals process were put in place. A Code of Practice was developed in collaboration with the Association of British Insurers, and in 1999 the government established the Genetics and Insurance Committee to evaluate tests and monitor compliance with principles of technical, clinical, and actuarial relevance.

From 2001 a five-year moratorium was voluntarily agreed, and subsequently extended, which restricted the use of test results below a certain insurance limit. As such, the HGAC further demonstrates the wider non-clinical context in which genetics was often discussed during this period and the tensions which might result. The Committee was set up to advise Ministers in the Department of Trade and Industry and the Department of Health, but its secretariat came from the Office of Science and Technology. There was therefore something of a 'tug of war' over its direction.[98] When, as described in the following chapter, a more powerful Human Genetics Commission was finally established in 2000, though a scientific and economic interest in genetics certainly continued and subsequently expanded, its positioning within the remit Department of Health demonstrated that a stronger clinical focus in the development of genetics policy had developed and that genetics as a whole had become more firmly established as an important health policy issue.

A Lonely Voice in the Wilderness: Genetics Meets Public Health

Many of the questions which had preoccupied clinicians and policymakers in Britain had also been identified by academics and practitioners with a population health perspective. If there was a foundational text for the field of Public Health Genetics then it was Muin Khoury's 1996 article 'From Genes to Public Health' in the *American Journal of Public Health*.[99] Having attended medical school in Lebanon during the 1970s and then trained in genetic epidemiology at Johns Hopkins School of Public Health, by the mid-1990s Khoury was well established at the US Centers for Disease Control and Prevention in Atlanta, working on congenital malformations. His thinking began to turn to the potential for bringing genetics and public health together more explicitly. The central premise of his 1996 article was that continuing advances in genetic technology – highlighted most clearly by the Human Genome Project – were likely to have profound implications for disease prevention. There was potential for paradigmatic change in conceptions of population health. In time, genetic susceptibility might come to be seen as a predictive and modifiable disease risk factor just as much as traditional environmental concerns.[100] Prevention might then follow at a behavioural, environmental, and clinical level. However, if this promise was to be realised then real thought would have to be given to how genetic advances were translated and applied.

What new population-based interventions might actually look like was far from clear. The same ethical and moral questions that were being debated elsewhere needed to be addressed. Potential safeguards envisioned by Khoury centred around consent and confidentiality, limits on the use of genetic tests in relation to insurance, comprehensive scientific evaluation of the ability of tests to identify underlying susceptible genotypes, and much wider education around genetics and its implications for medics and the public. Here Khoury felt that the 'public health community' was well placed to act, not just because the field itself would have to adapt to genetic developments, but because it represented a unique source of experience and had the ability to assess the safety, quality, and effectiveness of new developments in an appropriate way. His vision was central to the development of public health genetics.

Surprisingly, Khoury's article was initially rejected for publication because it was thought to have 'no interest to public health'.[101] He pressed the point with the editor of the *American Journal of Public Health* – 'at least you should have some discussion – this is THE journal' – and it was eventually published, alongside a comment piece by Gilbert Omenn, a leading epidemiologist and Dean of the School of Public Health at the University of Washington, which highlighted potential issues with the new approach in terms of managing informed consent. Nonetheless, Omenn agreed with Khoury's premise that genetics would grow in importance for disease prevention and that as a result public health had a 'special responsibility'.[102] As Khoury recalls, 'it was supposed to be a criticism of my paper, but I read it and I loved what Gil Omenn said, so instead of having one paper in the *American Journal of Public Health* we ended up having two papers ... that was the beginning of the field'.[103]

Having persuaded senior figures at CDC that there was real potential in this kind of approach, Khoury helped to establish and then chaired a task force which brought together different parts of the organisation and met with stakeholders, to establish what its role should be. The Director of CDC at the time was David Satcher – an experienced clinician administrator and future US Surgeon General – who had expertise in genetics and sickle cell disease, demonstrating the existing overlap between such genetic conditions and public health in the United States. The resulting strategic plan highlighted the importance of addressing ethical, legal, and social issues, training public health professionals, communication in genetics, and ensuring the quality of genetic tests.[104] It centred on the setting up of a CDC-wide office which would identify relevant scientific advances and help to integrate genetics into the CDC and wider public health. The Office of Public Health Genomics subsequently opened in 1997 with Khoury as Director. The Office was initially small with only a few members of staff, but its influence would be significant.

Among those inspired by Khoury's approach was Wylie Burke. Having met Khoury at a 1997 meeting arranged by the CDC and the National Human

Genome Research Institute on the discovery of the hemochromatosis gene, Burke began a sabbatical at the CDC in 1998, bringing an interest in the implications of genetics for routine clinical care to complement the wider interest in public health. The two were co-authors on a subsequent paper, alongside others with expertise in epidemiology, genetics, hepatology, molecular biology, public health and ELSI issues, which began to demonstrate the practical implications of their approach. The paper argued that despite the availability of genetic tests for hemochromatosis, issues including potential social harm and discrimination, meant that they should not yet be used in population screening. The discovery of a hemochromatosis gene did not instantly solve complex questions about the aetiology and molecular basis of the disease, and more data would be needed before genetic tests became more clinically appropriate than established tests using serum iron measures.[105]

Under the leadership of Omenn and the epidemiologist Melissa Austin, The University of Washington was also in the process of establishing an Institute for Public Health Genetics, which Burke joined when she returned from CDC. The importance of having a public health workforce which understood and was prepared for the impact of genomics, as the CDC had identified, saw the University of Washington introduce a Masters in Public Health Genetics in 1997, the first course of its kind, while the University of Michigan implemented a 'Public Health Genetics Interdepartmental Concentration' which brought together students from across different departments to study relevant issues.

These kinds of developments spoke to the emergence of a new field of 'Public Health Genetics', which was broad and interdependent with a number of other disciplines. Its immediate foundations were in genetic epidemiology and molecular biology, which were encompassed by the idea of human genome epidemiology – the 'application of epidemiologic methods and approaches in population-based studies of the impact of human genetic variation on health and disease' – which Khoury had helped to develop.[106] In 1998 a Human Genome Epidemiology Network – HuGENet – was established in the United States to promote collaboration and build a knowledge base. There was also overlap with other fields such as pathobiology, statistical genetics, bioinformatics, ecogenetics, pharmacogenetics, nutrition, health services, health policy, bioethics, and cultural anthropology. Only by appreciating overlapping trends in each of these fields would it be possible to fully appreciate the ways in which genomics might fundamentally impact conceptions of disease prevention. By the late 1990s therefore, there was a small but significant groundswell of opinion in the United States around the need to anticipate and prepare for such changes.

In Britain, there was, as yet, no groundswell, but complementary thinking was taking place. Ron Zimmern had become increasingly aware of the impact of genetics on clinical services, and thus the potential for improvements in his own field of public health. He saw the importance of ensuring that non-geneticists were part of conversations about the future of research, the development of genetics services and genetics policy. In 1997, Zimmern decided to make this his focus and formed the Public Health Genetics Unit (PHGU) in Cambridge. Well-placed support came from Richard Himsworth, Director of Research and Development at the Anglia and Oxford Regional Office of the NHS Executive and Professor of Health Research and Development at the University of Cambridge, and from Keith Peters, Regius Professor of Medicine.[107]

FIGURE 2.1   Dr Ron Zimmern

Source: Reproduced with the permission of the PHG Foundation.

As a result the PHGU, with Zimmern as Director, was formally established by the NHS Executive Regional office and became part of the University's Institute of Public Health, with its offices at the Strangeways Research Laboratory in Cambridge. Funding came from the NHS via the R&D division of the Regional office, and from Health Authorities in Bedfordshire, Cambridgeshire, Norfolk, and Suffolk.[108] This approach was unusual, but so was the concept of the PHGU. In its early years four key aims were set out:

> To keep abreast of developments in molecular and clinical genetics and in their ethical, legal, social and public health implications.
>
> To provide a link between academic research, clinical practice and the development of policy within the NHS for genetics and genetic services, including for the funding, development, staffing, organisation and provision of those services.
>
> To establish mechanisms for dialogue within the NHS between geneticists, physicians, public health and primary care professionals on matters related to genetics, molecular medicine and genetic services.
>
> To provide an epidemiological and public health perspective on NHS policy development for genetic and related services, including criteria for evaluating genetic testing and genetic screening programmes.[109]

Though this kind of work might be important, it was often theoretical rather than practical and would not be responding to an immediate clinical need. Given the way that genetics had developed, it was the kind of initiative which might previously have been taken up by a Regional Health Authority with the right outlook and the right kind of budget, but they had been replaced with eight NHS Executive Offices in 1996 as part of wider cost-saving measures. Individual local health authorities were unlikely to have such an interest or enough capacity, but Zimmern was able to use his connections to put the necessary pieces together.[110]

If Zimmern's vision was to be realised, the makeup of the PHGU team would be important. Carol Lyon, an experienced manager in the NHS and local government, was brought in as Business Manager. Alison Stewart, a trained biochemist, and Commissioning Editor of the scientific journal *Trends in Genetics* became Chief Knowledge Officer. Hilary Burton, Zimmern's public health consultant colleague, also joined the Unit, initially on a part-time basis alongside Health Authority work, before moving over full time. With only four members of staff, the PHGU was initially a small operation. Its areas of interest were also complex. As Stewart recalls:

> It was a very steep learning curve … to get up to speed with how health services were commissioned and organised, what public health was, and how it fitted into the whole picture, and then try and bring that together with genetics'.[111]

But it was a chance to do something unique. According to Burton, 'It was slightly scary in the sense that it was stepping out of the NHS/public career structure to some extent', but 'it got more and more interesting and more and more of an opportunity'.[112] The foundations for being able to do this were established in large part by building personal relationships and raising awareness of the Unit and its approach. Naomi Brecker, a Department of Health civil servant with responsibility for genetics policy during the late 1990s and early 2000s, recalls visiting Zimmern at his small office in Cambridge: 'He was always greatly enthusiastic in meeting people, networking people, bringing people together, a huge amount of energy in writing, speaking at events … to try and spread the word'.[113]

With this aim in mind Zimmern attended an event in Atlanta in May 1998, the first national conference on genetics and public health organised by the CDC – 'Translating Advances in Human Genetics into Disease Prevention and Health Promotion'. The keynote address was given by Gilbert Omenn, who spoke about the growing importance of genetics in public health and suggested that aspects of a wide range of other academic disciplines 'will become mainstream elements of academic public health and eventually of the practice of public health and preventive medicine'.[114] In his conference talk, Khoury discussed challenges and opportunities, stressing partnerships and coordination, ELSI issues, and training and education. The ELSI programme, academia, state departments of public health, federal agencies, and CDC, were all well represented, along with representatives of medical specialties with significance for public health such as child health, cancer, and chronic disease. The cross-cutting objectives of the conference were:

1. To update public health practitioners on developments in human genetics which can be incorporated into public health policy and practice.
2. To build partnerships between federal, state, academic, and private organizations to address challenges and opportunities for integrating genetics into disease prevention and health promotion efforts.
3. To assess significant public health issues, including ethical, legal, and social concerns associated with the integration of genetics into public health practice.[115]

The closing session of the conference saw confidence expressed that with 'effective public health leadership', 'the next millennium will see a new paradigm of individualized preventive medicine'.[116]

Strikingly though, Zimmern was the only delegate from outside North America. This would change in future years. But in 1998, though similar kinds of perspectives about the revolutionary potential of genetics, the need to think through difficult questions, and the need to ensure effective integration between the science of genetics and its clinical implications, were emerging in Britain, the threads had not yet been drawn together. That this subsequently happened was due at least in part to the work done by Zimmern and his colleagues, who increasingly highlighted these issues and emphasised the importance of public health perspectives, of the kind that had been developing in the United States. It was in Atlanta that Zimmern and Khoury first met. Their shared interests, demonstrated by the almost simultaneous setting up of the PHGU in England and the Office for Genomics and Public Health in the United States, quickly became apparent. Khoury reflects that 'I felt like a lonely voice in the wilderness … until I met my buddy Ron'.[117] The PHGU was soon seen as the 'twin' or 'kindred spirit' of the Office for Genomics and Public Health.[118] Other attendees at CDC conferences included Wylie Burke and Bartha Knoppers. They too made connections with Zimmern and could see, despite coming from different places and different disciplines, what their work had in common. According to Khoury. 'We never looked back. That was our introduction … all the activities that happened after that were as a result of that first contact'.[119]

Notes

1 E.T. Juengst, 'Anticipating the Ethical, Legal, and Social Implications of Human Genome Research: An Ongoing Experiment', *American Journal of Medical Genetics: Part A*, Vol. 185, No. 11, 2021, p. 3369.
2 A. Wolfe, 'Federal Policy Making for Biotechnology, Executive Branch, ELSI' in T.H. Murray and M.J. Mehlman (eds.), *Encyclopaedia of Ethical, Legal and Policy Issues in Biotechnology* (Wiley, 2000) p. 235.
3 Juengst, 'Anticipating the Ethical', p. 3373.

4 Interview with Dr Eric Meslin, January 2021.
5 A. Wolfe, 'Federal Policy Making for Biotechnology, Executive Branch, ELSI' in T.H. Murray and M.J. Mehlman (eds.), *Encyclopaedia of Ethical, Legal and Policy Issues in Biotechnology* (Wiley, 2000) p. 238.
6 E.J. Langfelder and E.T. Juengst, 'Ethical, Legal and Social Implications (ELSI) Program, National Center for Human Genome Research, National Institutes of Health', *Politics and the Life Sciences*, Vol. 12, No. 2, 1993, p. 274.
7 Interview with Professor Wylie Burke, November 2020.
8 Juengst, 'Anticipating the Ethical' p. 3371.
9 Interview with Dr Eric Meslin, January 2021.
10 T. Marteau and M. Richards (eds.), *The Troubled Helix: Social and Psychological Implications of the New Human Genetics* (Cambridge University Press, 1996). M.P.M. Richards, 'The New Genetics: Some Issues for Social Scientists', *Sociology of Health and Illness*, Vol. 15, No.5, 1993, p. 567–586.
11 House of Commons. Science and Technology Committee. *Human Genetics: The Science and its Consequences* (London: HMSO, 1995).
12 Interview with Dr Eric Meslin, January 2021.
13 Interview with Professor Wylie Burke, November 2020.
14 Ibid.
15 D. Bowen, A. McTiernan, W. Burke, D. Powers, J. Pruski, S. Durfy, J. Gralow, and K. Malone, 'Participation in Breast Cancer Risk Counselling Among Women With a Family History', *Cancer Epidemiology Biomarkers and Prevention*, Vol. 8, No. 7, 1999, p. 581–85. N.A. Press, Y. Yasui, S. Reynolds, S.J. Durfy, and W. Burke, 'Women's Interest in Genetic Testing for Breast Cancer Susceptibility may be Based on Unrealistic Expectations', *American Journal of Medical Genetics*, Vol. 99, No. 2, 2001, pp. 99–110.
16 V.A. McKusick, 'The Human Genome Organisation: History, Purposes and Membership', *Genomics*, Vol. 5, No. 2, 1989, p. 386.
17 B. Knoppers, A. Thorogood, and R. Chadwick, 'The Human Genome Organisation: Towards Next-Generation Ethics, *Genomic Medicine*, Vol. 5, No. 4, 2013.
18 Interview with Professor Bartha Knoppers, March 2021.
19 D.C. Werth and B.M. Knoppers, 'The HUGO Ethics Committee: Six Innovative Statements', *New Review of Bioethics*, Vol. 1, No. 1, 2003, p. 30. Interview with Bartha Professor Knoppers, March 2021.
20 http://hrlibrary.umn.edu/instree/geneticsresearch.html
21 HUGO Ethics Committee, 'HUGO Ethics Committee Statement on Benefit Sharing – April 9, 2000', *Clinical Genetics,* Vol. 58, No. 5, 2000, p. 366.
22 D.C. Werth and B.M. Knoppers, 'The HUGO Ethics Committee: Six Innovative Statements', *New Review of Bioethics*, Vol. 1, No. 1, 2003, p. 27.
23 Interview with Professor Bartha Knoppers, March 2021.
24 Ibid.
25 S. Harmon, 'The Significance of UNESCO's Universal Declaration on the Human Genome and Human Rights', *SCRIPTed*, Vol. 2, No. 1, 2005, p. 26.
26 Interview with Dr Eric Meslin, January 2001.
27 N. Lenoir, 'Universal Declaration on the Human Genome and Human Rights: The First Legal and Ethical Framework at the Global Level', *Columbia Human Rights Law Review*, Vol. 30, No. 3, 1999, pp. 537-587.
28 http://portal.unesco.org/en/ev.php-URL_ID=13177&URL_DO=DO_TOPIC&URL_SECTION=201.html
29 Ibid.
30 Ibid.
31 D. Wilson, *The Making of British Bioethics* (Manchester University Press, 2014).
32 Interview with Professor Sir Peter Harper, November 2020.
33 P. Nairne, 'The National Health Service: Reflections on a Changing Service', *British Medical Journal*, Vol. 296, May 28, 1988, p. 1519.

34 S. Lock, 'Towards a National Bioethics Committee', *British Medical Journal*, Vol. 300, May 5, 1990, p. 1149–50.
35 R. Chadwick and D. Wilson, 'The Emergence and Development of Bioethics in the UK', *Medical Law Review*, Vol. 26, No. 2, pp. 183–201.
36 Interview with Baroness O'Neill, January 2021.
37 Nuffield Council on Bioethics, *Genetic Screening – Ethical Issues* (London, 1993).
38 Wilson, *The Making*.
39 Nuffield Council on Bioethics, Genetic Screening – Ethical Issues (London, 1993) p. 93.
40 Wilson, *The Making*.
41 Interview with Professor Sir Peter Harper, November 2020.
42 *Report of the Committee on the Ethics of Gene Therapy*, Cm.1788 (London: HMSO, 1992). Hansard. House of Commons. 17 January 1992, Vol. 201. Col. 673.
43 *Report of the Committee on the Ethics of Gene Therapy*, Cm. 1788 (London: HMSO, 1992).
44 Hansard. House of Commons. January 17, 1992, Vol. 201. Col. 673.
45 L.A. Reynolds and E.M. Tansey (eds.), *Clinical Genetics in Britain: Origins and Development. Wellcome Witnesses to Twentieth Century Medicine*, Vol. 39 (London: Wellcome Trust Centre for the History of Medicine at UCL, 2010).
46 P. Brown, 'Gene Therapy Wins Official Blessing', *New Scientist*, No. 1805, January 25, 1992.
47 D.M. Geddes and N. Ross, 'Gene Therapy – Can We Afford It?', *Technology, Innovation and Society*, Vol. 14, No. 1, 1998, p. 11.
48 E.M. Jones and E.M. Tansey (eds.), *Medical Genetics: Development of Ethical Dimensions in Clinical Practice and Research, Wellcome Witnesses to Contemporary Medicine*, Vol. 57 (London: Queen Mary University of London).
49 *Population Needs and Genetic Services: An Outline Guide* (Genetics Interest Group, 1993).
50 Ibid.
51 *Report of a Working Party of the Standing Medical Advisory Committee on Sickle Cell, Thalassaemia and other Haemoglobinopathies* (London: HMSO, 1993).
52 G. Redhead, "A British Problem Affecting British People': Sickle Cell Anaemia, Medical Activism and Race in the National Health Service, 1975–1993', *Twentieth Century British History*, Vol. 32, No. 2, 2021, p. 203.
53 *Report of a Working Party of the Standing Medical Advisory Committee on Sickle Cell, Thalassaemia and other Haemoglobinopathies* (London: HMSO, 1993).
54 S.C. Davies, E. Cronin, M. Gill, P. Greengross, M. Hickman, and C. Normand, 'Screening for Sickle Cell Disease and Thalassemia: A Systematic Review With Supplementary Research', *Health Technology Assessment*, Vol. 4, No. 3, 2000.
55 D. Weatherall, *The New Genetics and Clinical Practice* (Oxford University Press, 1991). J.M. Connor and M.A. Ferguson-Smith, *Essential Medical Genetics* (Blackwell, 1991).
56 Interview with Professor Sir John Bell, August 2021.
57 Ibid.
58 *Realising our Potential: A Strategy for Science, Engineering and Technology*, Cm. 2250, (London: HMSO, 1993).
59 *Report of the Genetics Research Advisory Group: A First Report to the NHS Central Research and Development Committee on the New Genetics* (London: Department of Health, 1995) p. 1.
60 Peter Harper Interview with Professor Martin Bobrow, November 2004.
61 *Report of the Genetics Research Advisory Group: A First Report to the NHS Central Research and Development Committee on the New Genetics* (London: Department of Health, 1995) p. 14.
62 *Ibid* p. 16.
63 *The Genetics of Common Diseases: A Second Report to the NHS Central Research and Development Committee on the New Genetics* (London: Department of Health, 1995).

64 Ibid, p. 6.
65 Ibid., p. 4.
66 *Report of the Genetics Research Advisory Group: A First Report to the NHS Central Research and Development Committee on the New Genetics* (London: Department of Health, 1995) p. 1.
67 Peter Harper Interview with Professor Martin Bobrow, November 2004.
68 Interview with Professor Sir John Bell, August 2021.
69 Peter Harper Interview with Professor Martin Bobrow, November 2004.
70 Interview with Professor Sir John Bell, August 2021.
71 Welsh Health Planning Forum, *Genomics: The New Genetics on the NHS: The Cardiff Debate* (Cardiff, 1995) p. i.
72 Ibid., p. 5.
73 Ibid., p. 20.
74 Genetics Interest Group, *The Present Organisation of Genetics Services in the United Kingdom* (London, 1995) p. 9.
75 Ibid., p. 11.
76 Ibid., p. 15.
77 House of Commons. Science and Technology Committee. *Human Genetics: The Science and Its Consequences* (London: HSMO, 1995).
78 Hansard. HC Deb. 19 July 1996. Vol 281. cc1429.
79 House of Commons. Science and Technology Committee. *Human Genetics: The Science and Its Consequences* (London: HSMO, 1995).
80 Ibid., p. xvii.
81 Ibid., p. xxii.
82 Ibid., p. lxxiii.
83 Ibid., p. xxxi.
84 "Ibid.
85 Ibid., p. lvi.
86 Interview with Professor Sir Peter Harper, November 2020.
87 *Human Genetics: The Science and its Consequences*, Cmnd. 3061 (London: HMSO, 1996).
88 G. Vines, 'Gene Police Frozen Out', *New Scientist*, February 10, 1996.
89 Ibid.
90 Hansard. HC Deb. 19 July 1996. Vol. 281. cc1413.
91 *Human Genetics: The Government's Response*, Cmnd. 3306 (London: HMSO, 1996) p. 2.
92 Hansard. HC Deb. July 19, 1996. Vol. 281. cc1413.
93 Advisory Committee on Genetic Testing, *First Annual Report*, March 1998.
94 Interview with Professor Sir Peter Harper, November 2020.
95 C. Campbell, 'A Commission for the 21st Century', *The Modern Law Review*, Vol. 61, No. 5, 1998, p. 598–602.
96 *Human Genetics: The Government's Response*, Cmnd. 3306 (London: HMSO, 1996).
97 *The Implications of Genetic Testing for Insurance* (London: Office of Science and Technology, 1997).
98 Interview with Dr Mark Bale, February 2021.
99 M.J. Khoury, 'From Genes to Public Health: The Application of Genetic Technology in Disease Prevention', *American Journal of Public Health*, Vol. 86, No. 12, 1996, pp. 1717–722.
100 Ibid.
101 Interview with Dr Muin Khoury, November 2020.
102 Interview with Dr Muin Khoury, November 2020. G.S. Omenn, 'Comment: Genetics and Public Health', *American Journal of Public Health*, Vol. 86, No. 12, 1996, p. 1703.
103 Interview with Dr Muin Khoury, November 2020.

104 *Translating Advances in Human Genetics into Public Health Action: A Strategic Plan* (Centers for Disease Control and Prevention, 1997).
105 W. Burke, E. Thomson, M.J. Khoury, S.M. McDonnell, N. Press, P.C. Adams, J.C. Barton, E. Beutler, G. Brittenham, A. Buchanan, E.W. Clayton, M.E. Cogswell, E.M. Meslin, A.G. Moltulsky, L.W. Powell, E. Sigal, B.S. Wilfond, and F.S. Collins, 'Hereditary Hemochromatosis: Gene Discovery and Its Implications for Population-Based Screening', *Journal of the American Medical Association*, Vol. 280, No. 2, 1998, p. 172–78.
106 M.A. Austin, P.A. Peyser, and M.J. Khoury, 'The Interface of Genetics and Public Health: Research and Educational Challenges', *Annual Review of Public Health*, Vol. 21, 2000, p. 83.
107 Interview with Dr Ron Zimmern, November 2020.
108 *Genetics and Health: Policy Issues for Genetics Science and Their Implications for Health and Health* Service (Nuffield Trust, 2000).
109 Ibid.
110 Interview with Professor Sir Keith Peters, November 2020.
111 Interview with Alison Stewart, February 2021.
112 Interview with Dr Hilary Burton, January 2021.
113 Interview with Naomi Brecker, February 2021.
114 *1st Annual Conference on Genetics and Public Health: Translating Advances in Human Genetics into Disease Prevention and Health Promotion: Atlanta, Georgia, May 13–15, 1998* (Centers for Disease Control and Prevention, 1998) p. 39.
115 *Ibid.,* p. 5.
116 *Ibid.,* p. 85.
117 Interview with Dr Muin Khoury, November 2020.
118 Ibid.
119 Ibid.

Bibliography

1st Annual Conference on Genetics and Public Health: Translating Advances in Human Genetics into Disease Prevention and Health Promotion: Atlanta, Georgia, May 13–15, 1998 (Centers for Disease Control and Prevention, 1998).

Austin, M.A., Peyser, P.A., and Khoury, M.J., 'The Interface of Genetics and Public Health: Research and Educational Challenges', *Annual Review of Public Health*, Vol. 21, 2000.

Bowen, D., McTiernan, A., Burke, W., Powers, D., Pruski, J., Durfy, S., Gralow, J., and Malone, K., 'Participation in Breast Cancer Risk Counselling Among Women With a Family History', *Cancer Epidemiology Biomarkers and Prevention*, Vol. 8, No. 7, 1999.

Brown, P., 'Gene Therapy Wins Official Blessing', *New Scientist*, No. 1805, 25 January 1992.

Burke, W., Thomson, E., Khoury, M.J., McDonnell, S.M., Press, N., Adams, P.C., Barton, J.C., Beutler, E., Brittenham, G., Buchanan, A., Clayton, E.W. Cogswell, M.E., Meslin, E.M., Moltulsky, A.G., Powell, L.W., Sigal, E., Wilfond, B.S., and Collins, F.S., 'Hereditary Hemochromatosis: Gene Discovery and Its Implications for Population-Based Screening', *Journal of the American Medical Association*, Vol. 280, No. 2, 1998.

Campbell, C., 'A Commission for the 21st Century', *The Modern Law Review*, Vol. 61, No. 5, 1998.

Chadwick, R. and Wilson, D., 'The Emergence and Development of Bioethics in the UK', *Medical Law Review*, Vol. 26, No. 2, 2018.

Connor, J.M. and Ferguson-Smith, M.A., *Essential Medical Genetics* (Blackwell, 1991).

Davies, S.C., Cronin, E., Gill, M., Greengross, P., Hickman, M., and Normand, C., 'Screening for Sickle Cell Disease and Thalassemia: A Systematic Review With Supplementary Research', *Health Technology Assessment*, Vol. 4, No. 3, 2000.

Geddes, D.M. and Ross, N., 'Gene Therapy – Can We Afford It?', *Technology, Innovation and Society*, Vol. 14, No. 1, 1998.

Genetics and Health: Policy Issues for Genetics Science and Their Implications for Health and Health Service (Nuffield Trust, 2000).

Genetics Research Advisory Group: A First Report to the NHS Central Research and Development Committee on the New Genetics (London: Department of Health, 1995).

Genetic Screening – Ethical Issues (Nuffield Council on Bioethics, 1993).

Genomics: The New Genetics on the NHS: The Cardiff Debate (Welsh Health Planning Forum, 1995).

Harmon, S., 'The Significance of UNESCO's Universal Declaration on the Human Genome and Human Rights', *SCRIPTed*, Vol. 2, No. 1, 2005.

House of Commons. Science and Technology Committee, *Human Genetics: The Science and its Consequences* (London: HMSO, 1995).

HUGO Ethics Committee, 'HUGO Ethics Committee Statement on Benefit Sharing – April 9, 2000', *Clinical Genetics*, Vol. 58, No. 5, 2000.

Human Genetics: The Government's Response, Cmnd. 3306 (London: HMSO, 1996).

Jones, E.M. and Tansey, E.M. (eds.), *Medical Genetics: Development of Ethical Dimensions in Clinical Practice and Research, Wellcome Witnesses to Contemporary Medicine*, Vol. 57 (London: Queen Mary University of London, 2016).

Juengst, E.T., 'Anticipating the Ethical, Legal, and Social Implications of Human Genome Research: An Ongoing Experiment', *American Journal of Medical Genetics: Part A*, Vol. 185, No. 11, 2021, pp. 3369–3376.

Knoppers, B., Thorogood, A., and Chadwick, R., 'The Human Genome Organisation: Towards Next-Generation Ethics, *Genomic Medicine*, Vol. 5, No. 4, 2013.

Khoury, M.J., 'From Genes to Public Health: The Application of Genetic Technology in Disease Prevention', *American Journal of Public Health*, Vol. 86, No. 12, 1996.

Langfelder, E.J. and Juengst, E.T., 'Ethical, Legal and Social Implications (ELSI) Program, National Center for Human Genome Research, National Institutes of Health', *Politics and the Life Sciences*, Vol. 12, No. 2, 1993.

Lenoir, N., 'Universal Declaration on the Human Genome and Human Rights: The First Legal and Ethical Framework at the Global Level', *Columbia Human Rights Law Review*, Vol. 30, No. 3, 1999.

Lock, S., 'Towards a National Bioethics Committee', *British Medical Journal*, Vol. 300, May 5, 1990.

Marteau, T. and Richards, M. (eds.), *The Troubled Helix: Social and Psychological Implications of the New Human Genetics* (Cambridge University Press, 1996).

McKusick, V.A., 'The Human Genome Organisation: History, Purposes and Membership', *Genomics*, Vol. 5, No. 2, 1989.

Murray, T.H. and Mehlman, M.J. (eds.), *Encyclopaedia of Ethical, Legal and Policy Issues in Biotechnology* (Wiley, 2000).

Nairne, P., 'The National Health Service: Reflections on a Changing Service', *British Medical Journal*, Vol. 296, May 28, 1988.

Omenn, G.S., 'Comment: Genetics and Public Health', *American Journal of Public Health*, Vol. 86, No. 12, 1996,

Press, N.A., Yasui, Y., Reynolds, S., Durfy, S.J., and Burke, W., 'Women's Interest in Genetic Testing for Breast Cancer Susceptibility may be Based on Unrealistic Expectations', *American Journal of Medical Genetics*, Vol. 99, No. 2, 2001.

Realising our Potential: A Strategy for Science, Engineering and Technology, Cm. 2250, (London: HMSO, 1993).

Redhead, G., 'A British Problem Affecting British People': Sickle Cell Anaemia, Medical Activism and Race in the National Health Service, 1975–1993', *Twentieth Century British History*, Vol. 32, No. 2, 2021.

Report of a Working Party of the Standing Medical Advisory Committee on Sickle Cell, Thalassaemia and other Haemoglobinopathies (London: HMSO, 1993).

Report of the Committee on the Ethics of Gene Therapy, Cm.1788 (London: HMSO, 1992).

Reynolds, L.A. and Tansey, E.M. (eds.), *Clinical Genetics in Britain: Origins and development. Wellcome Witnesses to Twentieth Century Medicine*, Vol. 39 (London: Wellcome Trust Centre for the History of Medicine at UCL, 2010).

Richards, M.P.M., 'The New Genetics: Some Issues for Social Scientists', *Sociology of Health and Illness*, Vol. 15, No. 5, 1993.

The Genetics of Common Diseases: A Second Report to the NHS Central Research and Development Committee on the New Genetics (London: Department of Health, 1995).

The Implications of Genetic Testing for Insurance (London: Office of Science and Technology, 1997).

The Present Organisation of Genetics Services in the United Kingdom (Genetic Interest Group, 1995).

Translating Advances in Human Genetics into Public Health Action: A Strategic Plan (Centers for Disease Control and Prevention, 1997).

Vines, G., 'Gene Police Frozen Out', *New Scientist*, February 10, 1996.

Weatherall, D., *The New Genetics and Clinical Practice* (Oxford University Press, 1991).

Werth, D.C. and Knoppers, B.M., 'The HUGO Ethics Committee: Six Innovative Statements', *New Review of Bioethics*, Vol. 1, No. 1, 2003.

Wilson, D., *The Making of British Bioethics* (Manchester University Press, 2014).

3

PROGRESS

Spreading the Word

Having established the Public Health Genetics Unit (PHGU) in Cambridge in 1997, Ron Zimmern and colleagues set out to build interest in their work and share ideas. A personal and proactive approach came naturally to him. As a first step, he approached geneticists at each of the regional centres in Britain and travelled the country meeting them. 'Some of them were helpful', Zimmern reflects, 'and some of them were, not to put too strong a point on it, quite threatened, and didn't want to have anything to do with it'.[1] A number saw him as something of an 'interloper' as he had no formal genetics training and had taken it upon himself to enter the field. Among those who could see the value in what Zimmern and his colleagues were trying to achieve was Peter Harper in Cardiff. According to Harper, 'he learned very quickly, and he was apt at liaising and learning from people … he really proved a great asset once he had kind of got his feet under the table'.[2] Harper was also seeking to highlight the importance of genetics to those in public health. In Wales, he created a unique post of Consultant in Public Health Genomics in 1999 – jointly funded by the Welsh NHS and the University of Wales College of Medicine – with the aim of integrating genetic insights. The role was taken up by Layla Jader, a trained anaesthetist who had become interested in genetics whilst working with Harper on Cystic Fibrosis at the Institute of Medical Genetics in Cardiff before retraining in public health.[3] Among a range of activities, she organised courses in genetics and genomics for different groups of health professionals and policymakers.

In 1998, Zimmern created a Public Health Genetics Network to bring together those with a stake in the population health implications of genetics. This included public health practitioners from each health region in England, Scotland, Wales, and Northern Ireland, alongside a Department of Health civil servant, and

DOI: 10.1201/9781003221760-4

representatives of the MRC and the Wellcome Trust. The easiest to convince were those, like Harper, who had already been thinking about genetics from a population perspective. John Burn, Professor of Clinical Genetics at the University of Newcastle was also receptive. Yet, Burn suggests that at this point during the late 1990s Zimmern was still something of a 'lone voice'.[4]

What made Zimmern unconventional was not just his focus on public health but his networking skills and desire to involve the centre of government. The Department of Health took an interest in many areas of medicine – influencing their development and allowing their leading figures in turn to help influence policy – but up to now, there had been little interest in genetics at the centre, which had rarely taken the initiative over service development or clinical practice. Although this was beginning to change by the late 1990s, most practitioners still thought it more important to be part of debates taking place in and around the Royal Colleges and leading professional societies. Zimmern however wanted to work with the Department of Health. There were others in academia and the NHS interested in the questions the PHGU was raising. Key figures such as Burn appeared to have a foot in each camp. But the unique position of the PHGU, engaging not just with academia, clinicians, and practitioners, but with policy-makers, while independent of each, afforded Zimmern and his colleagues a valuable degree of flexibility and freedom of thought and action. They were able to develop a particular set of ideas and pursue them.

Organisations such as the Clinical Genetics Society – which had played an important role in the development of professional training – continued to do valuable work. Its Lead Clinicians Group, for example, allowed individuals to share best practice around how to engage with the representatives of local authorities who now 'commissioned' genetics services under the 'internal market'. According to Frances Flinter, Professor of Clinical Genetics at King's College London and Consultant Geneticist at Guy's and St Thomas' NHS Foundation Trust, 'We knew our commissioners very well, we knew how many sugars they had in their tea. There were people in other departments who'd never met their commissioners'.[5]

The formation of the British Society for Human Genetics (BSHG) in 1996 was another important development. This new umbrella organisation brought together the Clinical Genetics Society with the Association of Clinical Cytogeneticists, the Clinical Molecular Genetics Society, and the Genetic Nurses and Social Workers Association. There had been a growing sense of the need to offset some of the diversification and fragmentation in genetics and present more of a united front, including when engaging with health policymakers. Peter Farndon, Professor of Clinical Genetics at the University of Birmingham, was a leading figure in the formation of the BSHG, alongside Martin Bobrow, Professor of Medical Genetics at the University of Cambridge, and Andrew Read, Professor of Human Genetics at the University of Manchester.

The BSHG gained representation on a number of important national committees and made a 'large noise' but with limited initial impact.[6] Burn, a later

Chair, reflects that in its early years 'we would write things for the Department of Health, but … the way the structure of things was then organised, the NHS and the Royal Colleges were the kind of power base, and we didn't really get through'.[7] 'Politicians were fairly oblivious to us in the nineties' he suggests.' We were kind of knocking on the door, but we weren't really making a big impact'. Only with the development of new and expensive treatments for a number of rare diseases during the 2000s, did this really begin to change.

The issue that perhaps helped to raise the profile of genetics the most was stem cell research. In 2000, Liam Donaldson, the Chief Medical Officer (CMO) between 1998 and 2010, chaired an Expert Advisory Group on Therapeutic Cloning which produced the report *Stem Cell Research: Medical Progress With Responsibility*. It argued for a change in the law to allow stem cell research for therapeutic purposes to be carried out on embryos created for that purpose.[8] Burn was a member of the group, as were John Polkinghorne, David Weatherall, and Dian Donnai, a Consultant Clinical Geneticist in Manchester who became the Consultant Adviser to the CMO on Genetics in 1998. MPs were given a 'free vote' on the required legislative changes. The law was changed, but the key parliamentary debate in December 2000 often returned to the perceived moral implications of cloning. There were also clinical questions. The Genetic Interest Group and others supported the changes because they offered the prospect of future treatments for those suffering from inherited conditions. There were also economic questions. Stem cell research was seen as an important part of the future development of the biotechnology industry in Britain.[9]

Awareness of genetics was gradually spreading from scientific and clinical circles into wider debates about social and economic policy. In 1998, the left-of-centre Institute for Public Policy Research [IPPR] published *Brave New NHS? The Impact of the New Genetics on the Health Service*.[10] The author was Jo Lenaghan, a Research Fellow in Health Policy at IPPR who went on to become a Special Advisor to the NHS Chief Executive and Director of Strategy at Health Education England. Research had also been carried out by IPPR's Liz Kendall, who went on to be a special advisor to Patricia Hewitt, Secretary of State for Health between 2005 and 2007. Kendall is now a Labour MP. Familiar names on the study's advisory group included Hilary Harris and Alistair Kent, while others, including Zimmern, read early drafts of the report. *Brave New NHS* outlined what were now, at least to geneticists, familiar arguments about the potential for genetics to fundamentally change the nature of health care and the need to think through a series of important, linked, issues.

The report's key findings were couched in terms of what genetics would mean for the National Health Service. The complexities and uncertainties were acknowledged. The report identified differences of emphasis amongst what were described as 'enthusiasts' and 'more cautious commentators' around a shift in focus from single-gene disorders to common complex conditions.[11] It acknowledged the challenges of innovation: 'It is impossible to predict which future will prevail. It is likely that development will occur at different rates for different diseases with different clinical implications'.[12] Even so, the report recommended that the starting point for policymakers should be a recognition that the NHS was largely set up to 'diagnose and treat' illness rather than 'predict and prevent' it. Its focus was predominantly phenotypic rather than genotypic. As a result, the development of services had often lacked coherence and coordination. The development of new genetic tests and screening programmes would present significant challenges in terms of funding, organisation, and ensuring clinical and cost effectiveness. *Brave New NHS* identified a need for a new set of guiding principles for the NHS. An understated but recurring theme in the report was the importance of public health to many of these debates. 'Public Health Genetics' as a term was not used, and public health was not necessarily seen as having a unique contribution to make in the way had been suggested by Muin Khoury for example. But the report did pick up on a number of points that had been made by Harper, Zimmern, and others, and identified an important disconnect between public health and clinical genetics:

> For example, if a woman received ante-natal screening for cystic fibrosis, which proved to be positive, and following counselling decided to go ahead with a pregnancy, how would purchasers measure this as an outcome? In this instance, the NHS will have enabled one individual to make an informed choice about their life, which a Consultant Clinical Geneticist would regard as a success – but what about a Director of Public Health working for a cash-strapped health authority?[13]

If the potential of genetic medicine was to be realised then a new kind of multidisciplinary approach with new mechanisms for sharing and integrating expertise within the NHS would have to be found – an issue that deeply concerned those in and around public health genetics.

Complementary ideas had also been put forward in a series of reports produced by the Clinical Genetics Committee of the Royal College of Physicians. In 1996 *Clinical Genetic Services into the 21st Century* had described how primary care and other clinical specialties might change and take on new roles as genetics influenced understanding of common diseases, leaving clinical geneticists to handle more complex cases.[14] One means of facilitating this would be joint appointments across specialties. In 1998, the report *Clinical Genetic Services: Activity, Outcome, Effectiveness and Quality* made the case for a new series of quality indicators around the scope, accessibility, responsiveness and audit of genetic services.[15] This was followed by *Commissioning Clinical Genetic Services* which cited cancer genetics as a field in which referrals to specialist services were rising and recognised that greater 'filtering' was needed in order to identify which cases could be handled by primary care and other specialties.[16]

This issue had also been considered by a working group established by Donaldson as CMO, and chaired by Harper, which recommended an integrated three-tier structure for the organisation of cancer genetics services. It encompassed appropriate referrals from primary care, cancer units staffed by oncologists with an understanding of genetics for individuals with moderately increased risk and specialist services led by consultants trained in both oncology and genetics for the most high-risk patients.[17] Aside from highlighting the importance of adequate staffing, training, facilities, time spent in the clinic and managerial support, *Commissioning Genetics Services* repeated the case for the commissioning of services on a regional basis. While this had largely continued on the ground, there was increasing recognition of the dislocation and fragmentation in the development of specialist services that the NHS internal market was causing. A specific proposal was for each new 'regional commissioning group' to include at least one public health consultant with experience of genetic services.[18]

The Royal College of General Practitioners had also established a Faculty Genetics Group in the North West of England which recognised that primary care would be central to the success of genetics in medicine. As its report *Genetics in Primary Care* described, 'We stand on the threshold of a dramatic expansion of our ability to influence favourably the population's health'.[19] The focus was still on initial issues such as training and education and collaboration with other specialist groups, and the need to consider priority setting, but it was clearly seen that:

> Medical genetics is in fact the quintessential arena for shared care. The technology and the concepts of probability and risk are sufficiently complex to demand intermittent specialist involvement. Yet only primary [care] can cater for the life time needs of patients and families with chronic genetic disabilities and with threat of future disease in family.[20]

An important theme throughout these reports was the need to ensure effective integration and coordination between specialist genetics services and the treatment of inherited disorders and clinical practice as a whole and the treatment of common complex conditions. The importance of having a wider multidisciplinary view which included specialties such as public health and primary care was now increasingly well understood.

The most detailed analysis of the place of genetics in health care during this period came from the PHGU. In 1997 the Nuffield Trust, a leading health policy think tank, had launched a *Policy Future for UK Health* programme of study which sought to think 'strategically about the issues that were likely to determine the future of health in the United Kingdom and the action that would have to be taken to achieve what were decided upon as appropriate healthcare goals'.[21] A range of issues was addressed, including science and technology, patient engagement, and finance. Advances in genetics and the implications for health services were not initially considered by the Nuffield Trust as the subject was thought to be too complex. There was an opportunity to provide this kind of analysis therefore, something which the PHGU was well placed to do. Zimmern put together a proposal for John Wyn Owen, then Secretary of the Nuffield Trust.

The PHGU was commissioned to undertake what became known as the 'Genetics Scenario Project'. Zimmern reflects that this was 'our first big project, our first big break' – the thing which 'really put us on the map'.[22] There was also collaboration with the Genetic Interest Group.

The central aim of the Genetics Scenario Project was 'to assess the impact of advances in genetics and molecular biology on the organisation, funding and provision of clinical services, on changes in clinical practice, and on the potential for disease prevention and public health action'.[23] Although the funding was limited, Owen ensured that the Nuffield Trust provided practical and institutional support, as well as 'very good wines' at working dinners at its central London offices. At the heart of the project were eight workshops with groups of key stakeholders held throughout 1999. The lists of participants read like a 'who's who' of genetics and health policy in Britain. The workshops' membership was drawn from the pharmaceutical industry; physicians, public health physicians, and general practitioners; social scientists and ethicists; policy commentators; policy-makers within government and the NHS; medical geneticists; physicians with expertise and experience in genetics; and patient representatives. Three main questions were asked at each workshop:

I. What do you think are likely to be the key issues regarding the use of emerging genetic technology to improve health?
II. What are the principal uncertainties in the immediate future that must be identified in order to develop health policies in relation to the new genetics?
III. What are your principal recommendations to government ministers about the development of future health policy as it shapes developments in medical genetics in respect of first, national policy priorities, and second, the organisation, funding, and provision of clinical services within the NHS?[24]

The questions were used to help frame discussions and draw out what the expert groups perceived to be the most important issues, the key values at stake and the necessary drivers for change. The eight workshops were then followed by a two-day policy development gathering in November 1999, with one or two representatives of each group and a small core group of invitees. Here, all the material that had been gathered was brought together and analysed to identify common themes.

Alongside Zimmern and Hilary Burton from the PHGU, the project team included Max Lehman from the Nuffield Trust, Peter Wightman from Cambridge and Huntingdon Health Authority, and Tom Ling, a research consultant and academic at Anglia Ruskin University. Tony Hodgson of Idon Associates, with whom Zimmern had worked before, was also influential. He and Ling had experience of 'scenario planning', the methodology which was used to help identify key issues and develop long-term plans. The 'ideal' scenario was one in which all the benefits of genetics were realised with none of the potential drawbacks. Different scenarios were put to the workshop participants in order to gauge their responses. They then used the 'Hexagon Method' – 'a kind of organised "brainstorming" that encourages people to escape conventional linear ways of thinking about problems' – which centred on writing key points on little hexagons with 'the ability to arrange and to rearrange the "hexagons" ... to provoke new ways of thinking about familiar issues'.[25] The final report, *Genetics and Health: Policy Issues for Genetic Science and Their Implications for Health and Health Services*, was published in 2000. Christopher Cook, a freelance journalist and broadcaster, was brought in to co-author the report with Zimmern, highlighting the mainstream nature of the intended audience. One of the central points that *Genetics and Health* sought to get across was that debates were ongoing and there was still plenty of uncertainty. As John Wyn Owen noted in his Foreword:

> No one can predict exactly what the impact of genetics on medical practice and health care is likely to be or when it will happen. Nevertheless, on the available evidence its significance cannot be underestimated, and we need to plan strategically for the kinds of changes that undoubtedly lie ahead.[26]

The report likened the impact of genetics to 'a tidal wave, a tsunami, sweeping all before it as it bursts upon the shore', and argued that 'it is incumbent upon us to prepare for the tides that will eventually change the shape of the shoreline of the nation's health'.[27] Although the report acknowledged it was not possible to accurately predict the pace of these changes, it noted that advances in genetics would have a significant impact on the organisation and funding of health services. The ability to make informed and systematic judgements about that organisation and funding would be crucial. With this in mind, the report was aimed directly at senior policymakers and set out proposals for a new national 'policy framework'.

In addition to 'General Drivers' such as advances in IT, globalisation, and demographic change, the project identified two 'Specific Drivers' – public attitudes towards genetics and science and its capacity to improve human health. There were six key policy areas in which decisions would shape the impact of those drivers: regulation, education, finance, information and confidentiality, commercial considerations, and the science base. Service delivery would be influenced by provision in terms of testing, screening, and counselling, and wider ethical legal and social issues such as the use of genetic information and human rights. This could all be neatly conceptualised in an interconnected hexagon diagram.

The national policy framework would also need to be underpinned by an agreed set of values: equity, effectiveness, affordability, quality, partnership, transparency, and integrity.

This methodological approach led to a series of recommendations. If the benefits of genetics were to be realised *Genetics and Health* argued, then a regulatory framework would be required that was 'light enough to encourage promising developments in genetics but rigorous enough to protect the public'.[28] It proposed the development and implementation of a strategy for genetic literacy, including medical school curriculum changes, public campaigns and a 'systematic approach to the training and education of health professionals'.[29] A set of

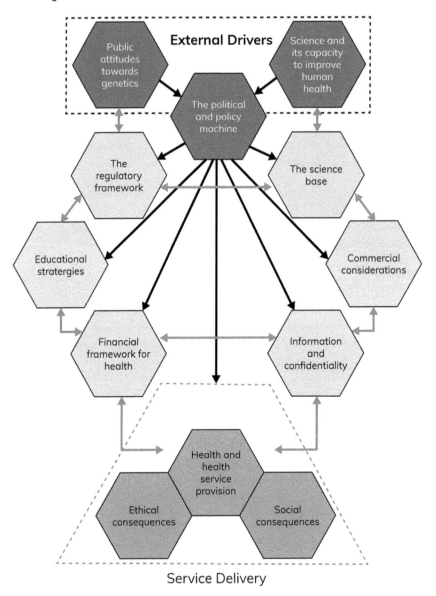

FIGURE 3.1 Specific Drivers for Genetics from *Genetics and Health: Policy Issues for Genetics Science and Their Implications for Health and Health Service* (Nuffield Trust, 2000). Reproduced with the permission of the Nuffield Trust

principles around the use of genetic data which centred on the public interest and informed consent would need to be developed. In terms of finance, it was argued that 'The UK should have regard to the cost pressures posed by the impact of genetic technologies on the health service and ensure by whatever means the necessary resources for the funding of cost-effective interventions arising out of

genetic science'.[30] On the commercial side, the aim should be to 'foster an entrepreneurial spirit within the biotechnological and pharmaceutical industries' and
'stimulate partnerships between them and academia'.[31] Alongside this however
there would need to be a 'commitment to commercial integrity' around access to
and the cost of drugs.

The report also called for further investment in scientific infrastructure and a
rise in the status of science and scientists, in parallel with a provisional five-year
strategy for the regional genetics services. This would need to include assessments
of the validity of genetic tests and their social and ethical implications, new mechanisms for commissioning, and reviews of the responsibilities and recruitment of
geneticists. There should also be a 'horizon scanning mechanism to monitor future
developments particularly in pharmacogenetics and in the use of genetic tests of
susceptibility as a means of predicting and preventing disease'.[32]

The recommendations in *Genetics and Health* were relatively general, and the
authors conceded that much of the report was 'conjectural'. There was no further
detail on the potentially difficult choices that future policymakers might face if
adequate funding for genetic services were to be maintained, for example. But
while most of the ideas were not new, this was the first time that they had been
brought together in such a systematic way, and the first time in which such
breadth of expert advice had been drawn upon in an attempt to plan for the future.
The degree of coordination and cooperation that would be necessary was clear.
There was an inherent recognition that the success of such plans would be closely
linked to the mechanisms of health service delivery, which would mean adapting
the existing structures and approaches of the NHS. There would also have to be a
significant partnering role for the biotechnological and pharmaceutical industries.
Commercial interests could not be ignored but they could be incorporated.
Genetics was taken to have a broad meaning which meant that there was a broad
view of the potential benefits, but it was also made clear that environmental determinants and other factors would continue to be of 'inestimable importance'.[33]
The scope and shape of this vision was therefore critical. As Zimmern reflected,
'Why I am really proud of this is that we laid out and identified all the issues, and
now twenty-five years later, when we have got a strategy, we haven't missed out
on any of the issues'.[34] While its wide-ranging approach and the broad timescales
involved make it difficult to judge the direct influence of the Genetics Scenario
Project, a number of issues, including professional education, technological infrastructure, and the coordination of services, subsequently became the focus of
significant policy discussions.

PHGU's next innovation was to establish a Genetics and Health Policy course –
the first of its kind – run explicitly for senior policymakers. The first cohort in 2000
included Consultants and Specialist Registrars in Public Health Medicine,
Department of Health civil servants, a Health Authority Chief Executive, as well
as other senior NHS managers and commissioners and a number of academics.
The residential course at the Genome Campus in Hinxton ran for five days.
Lectures were delivered by experts from a range of disciplines and covered topics

such as the Human Genome Project, pharmacogenetics, and UK Biobank – which is discussed in more detail below. Other sessions considered stem cell medicine, gene patenting, and the ethical issues associated with genetic testing and genetic information. The course was wound up in 2006 because it had come to take up too much time and resources. But it had proved an effective mechanism for disseminating PHGU ideas and approaches. Each participant was sent away with 'the most enormous kind of reference library with everything there was to know at the time about genomics and public health'.[35] The PHGU's later assessment was that the courses had 'created a cadre of senior professionals with the knowledge and confidence to take a leadership role on genetics in their own field'.[36] Zimmern personally found delivering the courses 'wonderful fun'.[37] There were limits to the scope of their action, and public health genetics itself remained a relatively niche area of interest, but the PHGU and a number of like-minded actors were beginning to make some headway during this period.

Mechanisms of Change

During the late 1990s and early 2000s, Zimmern continued to build his personal connections. For example, he invited Sheila Adam, Deputy Chief Medical Officer at the Department of Health, to visit him at his office in Cambridge. The two remembered each other from their time as public health trainees. Adam had since become a public health consultant and then Director of Public Health at North West Thames Regional Health Authority during the 1980s, before joining the NHS Executive as Head of Mental Health and NHS Community Care in 1995. She became Deputy Director of Health Services in 1997, and then Director and Deputy CMO in 1999. As such, she was in an important position and able to feed into policy discussions. During their meeting, Zimmern emphasised the scale of the changes that were likely to result from advances in genetics, and the need for the Department of Health to understand and be well prepared for them. He told Adam that 'we can't have policymaking without genetics being a large part of it'.[38] Zimmern thought that Adam was receptive to his ideas but, given the scope of her responsibilities and the many pressures on senior civil servants' time and attention, he was unsure about the impact his words would have.

In fact, Adam was already conscious of the growing significance of genetics in health care from her time at North West Thames, where Robert Williamson, then Professor of Molecular Biology and Genetics at St Mary's Hospital in London, led important research around gene mapping that included the identification of the genes associated with cystic fibrosis and Duchenne muscular dystrophy.[39] She also knew that a small number of Department of Health civil servants working in other policy divisions had relevant knowledge about genetics and was surprised that it had not yet become a more concerted area of activity. She subsequently took steps to try and ensure more joined-up thinking in the development of genetics services, including between the different sides of the Department that were responsible for policy and delivery, as had been achieved in other areas such as mental health.

In 1998, short-term funding was brought together in order to create a new genetics post to which the experienced civil servant Naomi Brecker was appointed. Brecker had joined the Department of Health having initially trained in public health. Because the role was new she had something of a blank slate to work with and was able to talk to key people in the field, both inside and outside the NHS, and get across the important issues. Brecker knew that Zimmern had been an influential Director of Public Health, and he was one of those with whom she was advised to meet. As she recalls:

> I remember spending a few hours with him … he didn't just want it to be a little add-on on the margins of Cambridge. He was very adept at trying to get it integrated into more mainstream thinking. So, our paths kept crossing.[40]

Over the next few years, Brecker attended a number of conferences and workshops organised by the PHGU as part of her role and was a member of the Public Health Genetics Network. A small Genetics Policy Unit was subsequently created inside the Department of Health, with the aim of facilitating collaboration and coordination, making sure that the NHS was able to realise the potential benefits of genetics and that the inherent social and ethical questions were addressed.[41] There were also the everyday tasks of briefing Ministers and responding to questions asked about genetics in the House of Commons. Brecker recalls:

> My role at the Department of Health was to act as more of a bridge … to bring together lots of expertise and then to try and facilitate the conversations they needed to have with the policy and power makers, who could perhaps find some resources or find the entrée into sort of a political awareness, so that you could actually have the system changes.[42]

Brecker was soon seen as the 'fount of all wisdom' by colleagues.[43] In 2001, an internal reorganisation created a new Genetics Branch within the Clinical Quality Ethics and Genetics Directorate of the Public Health Group, which dealt with issues such as human tissue, sexual health, and abortion. Brecker became its Medical Adviser. The focus was now more clearly on the organised development of specialised genetic services, which were growing in importance. There was also a recognition that the impact of genetics on common diseases and the resulting increase in clinical burden meant there would have to be a strategic view of service delivery. The Department of Health was beginning to address these kinds of issues seriously and a more formal policy infrastructure developed as a result.

In July 1999, Zimmern was invited to give a presentation to the NHS Executive Board – an acknowledgement of his persistent attempts to connect with high-level policymakers. The Permanent Secretary at the Department of Health between 1997 and 2000, was Christopher Kelly. He and Zimmern had been contemporaries at Trinity College, Cambridge, and Zimmern had also worked for

Kelly's father, a distinguished neurologist, whilst a registrar at the National Hospital in London. Zimmern's presentation set out the 'basics' of genetics and the determinants of health, including the balance of gene-environment interactions in relation to different conditions. For example, the development of Duchene muscular dystrophy was 'Totally Genetic' while being struck by lightning was 'Totally Environmental'. He described the ways in which the Human Genome Project would facilitate a better understanding of disease mechanisms and open up opportunities for new therapies and prevention. He emphasised the need for partnerships and interdisciplinarity in policymaking, considering things such as ethical, legal, and social implications. He identified specific issues that should be addressed in the shorter term such as genetic literacy and professional training and strategies for the development of laboratory services, screening, and testing.

As before, none of these issues was new but the level at which they were now being discussed was significant. Yet although awareness was increasing, it was difficult to know how this kind of engagement would actually translate into policy development. As Zimmern reflects, 'It was exciting to be asked to address people right at the top of the NHS tree, but having addressed them, what happened ... I can't tell you'.[44] Even so, the network building and relatively top-down approach adopted by the PHGU had begun to pay dividends. Civil servants knew that the Unit could be asked for informal advice, was likely to offer a useful perspective, and could provide experts to sit on relevant advisory groups, working parties, and committees.

An early manifestation of this new kind of direction was an informal 'Stakeholder Group' which met at the Department of Health. Mark Bale, a key civil servant who joined the Department from the Health and Safety Executive in 1999, recalls that 'there was no kind of paperwork, it wasn't a committee ... it was whoever was ready and willing to drink DH coffee and have a chat'.[45] Zimmern became a member. A joint working group for the NHS Executive and the Human Genetics Commission, chaired by Martin Bobrow, was also responsible for driving some important developments. Their report *Laboratory Services for Genetics* published in 2000 argued that while the informal national network of genetic testing had largely worked well, a new national strategy was needed to make sure that tests for rare genetic conditions were widely available, that the benefits of new tests were fully evaluated prior to introduction, and that they were equitably delivered across the country.[46]

The report laid the foundations for much of the new investment in genetics – particularly in terms of state-of-the-art biomedical laboratory equipment – which followed. A Genetics Commissioning Advisory Group [GenCAG] was also established in 2000 as a sub-group of the National Specialist Commissioning Advisory Group, which advised Ministers about the commissioning of services for rare and expensive to treat conditions. It brought together a range of expert voices and ensured greater national coordination of genetic services.[47] Guidance was provided to the NHS purchasers of genetics services, particularly around the planning of resources and manpower. A set of 'quality markers' was developed and adopted.[48] The Group was chaired by Sir John Pattison, the Director of Research and Development in the Department of Health between 1999 and 2004. Zimmern was a member, alongside representatives of professional bodies, commissioners, and the Genetic Interest Group.

The National Screening Committee – first created in 1996, and chaired between 1998 and 2006 by Henrietta Campbell, the Chief Medical Officer for Northern Ireland – was also becoming an important venue for related policy debates. It was tasked with balancing the benefits and risks of new screening programmes and establishing whether they should be funded by the NHS. The focus was not initially on genetics, and clinical geneticists were not specifically represented, but the continuing advancement of new technologies in relation to a wider range of conditions meant that genetic questions became increasingly important. The National Screening Committee took a 'cautious and critical' approach, setting relatively strict criteria for the introduction and evaluation of screening programmes.[49] Another body which began to facilitate related changes was the Joint Committee on Medical Genetics. Established in 1998 and led by the Royal College of Physicians, the Royal College of Pathologists, and the British Society for Human Genetics, it represented a further attempt to consolidate the voice of a range of professional groups. The Joint Committee also had well-established links with the PHGU. It sought to present a collective view of the development of genetics services and communicate this to the Department of Health and other organisations, in effect serving as the profession's main advisory body. Naomi Brecker attended meetings as a representative of the Department of Health. Peter Farndon, the Joint Committee's first Chair, reflects that:

> The great thing about the committee was if the committee decided something with everybody, all the different parties agreeing, we had a mechanism to make it work. If it was education we could feed it through the Royal Colleges. If it was a service issue we could feed it through the BSHG into the Genetics Service. If it was policy issues we could ask the Department of Health what they thought about them. I think that the JCMG was the right thing at the right time.[50]

The most high-profile example of change in the advisory and regulatory environment during this period, however, was the establishment of the Human Genetics Commission (HGC) in 2000.

A more conducive political atmosphere for genetics had emerged following the election of a Labour government in 1997. The case for having an independent statutory body like the HGC, which had once concerned senior officials, was now more readily accepted, and in time genetics came to be seen as a more positive proposition – an area of policy which provided an opportunity to talk about proactively about the future of health and health care – something which naturally appealed to Ministers. The HGC's role was to advise about developments and identify priorities in research and service delivery. Another important element was to encourage debates and consult the public and wider stakeholders. The HGC subsumed the existing Human Genetics Advisory Committee, Advisory Committee on Genetic Testing, and Advisory Group on Scientific Advances in Genetics. Calls for such a body which had been growing for a number of years, including by the Nuffield Council on Bioethics and the House of Commons Science and Technology Committee, were finally recognised.

The pressures which ultimately produced the HGC were in part technological – reflecting the need for policymakers to understand complex advances in biotechnology and pharmaceuticals – and in part social and political. During the mid-1990s a number of high-profile stories, including the birth of Dolly the Sheep via cloning, the spread of Bovine spongiform encephalopathy (BSE) and its links to Creutzfeldt–Jakob disease in humans, and the development of genetically modified foods, had coloured popular perceptions of science. An official review in 1999 identified restoring public trust as an important issue.[51] The establishment of the HGC demonstrated that there was a degree of separation between human genetics and wider scientific and genetic issues. An Agriculture and Environment Biotechnology Commission and the Food Standards Agency were also established in 2000, in what *Nature* described as an outbreak of 'committee-mania'.[52] There was also a longer-term context in terms of a lingering fear of eugenics and genetic determination. Baroness O'Neil suggests that such concerns were 'always around in the undergrowth'.[53]

The first Chair of the HGC was the barrister and broadcaster Baroness Kennedy. Peter Harper, a member of the Commission from 2000, nominally as the representative of the Welsh Chief Medical Officer, was among those who look back on her role positively: 'She was very influential, both in terms of all her contacts within parliament, but also in terms of her ability to get things across to the public'.[54] The focus on ethical, legal, and social issues and the open and upfront way in which the HGC worked, getting out ahead of problems before harmful applications might develop and helping to disentangle medical genetics from wider controversies, also aided the development of the field.[55] The HGC had a broad basis, including representatives of clinical geneticists, genetic counsellors, nurses, lawyers, bioethicists, and the pharmaceutical industry. Alastair Kent of the Genetic Interest Group was first appointed to the Commission in 2002 and helped to provide the voice of patients. He felt that:

> It demonstrated independent thought. That meant that when people looked at the advice that was given to government they could see that it wasn't partial, and it wasn't influenced by … powerful vested interests with deep pockets. I think the principal contribution that we made was the creation of a climate of respect, of public trust.[56]

Early issues considered included attitudes to personal genetic information, the use of genetic tests in insurance, and preimplantation genetic diagnosis.[57] The latter was also the subject of a unique public consultation undertaken jointly by the

HGC and the HFEA in 2003.[58] The decision to recommend a moratorium on genetic information in the setting of insurance premiums, something which subsequently came into effect in 2001, was one of the HGC first achievements.[59]

Influential reports subsequently examined direct-to-consumer genetic tests and the controversial idea of establishing a national DNA database.[60] A visit to the Forensic Science Service laboratory in Birmingham had alerted HGC members to the lack of an ethical underpinning for the retention of genetic information as part of the database, something which was subsequently rectified.[61] Yet, the focus on ethical, legal, and social issues could sometimes be a source of frustration for clinicians who were also anxious to discuss the development of genetic services. As Bruce Ponder reflects, for example:

> I mean I am not very good on committees anyway because I tend to keep quiet unless I am asked, but I had a sense that the undoubtedly important ethical, legal, and social issues were dominating over what the scientists saw as the practical issues of how you actually got genetic services going.[62]

According to Mark Bale, who became the Secretary of the HGC in 2000:

> The first meeting had not gone brilliantly, I think there was a kind of sense of some of the people weren't quite sure why they were there ... bit of a hostility between the Chair and some of the industry people ... there was a sense that there are some big ethical issues here that need to be carefully explored.[63]

There were also organisational problems. The Commission's second meeting was postponed by the widespread fuel protests in 2000, which meant that the imperative to move around the country and engage with the public was somewhat delayed. Bale reflects:

> I felt I had joined a committee, and the first thing that we did was only ever meet as a very small working group ... I think there were some people who thought it was a bit of a brave experiment, and it would be better to go back to the old ways of working.[64]

Around the HGC therefore, which suited the 'professional talkers' well – the philosophers and the lawyers – there was still a space that Zimmern, the PHGU and others were able to occupy, bringing an interest in related issues but tying them in more directly with practical considerations.[65]

Having addressed the NHS Executive Board in 1999, the following year Zimmern was invited to give a presentation at a seminar on genetics hosted by the Secretary of State for Health, Alan Milburn. His experience told him that genetics had now 'got to the top of the pile'.[66] Representatives of patient groups, clinicians, scientists, and the pharmaceutical industry also attended. Zimmern emphasised many of the same messages as before, drawing on the Genetics Scenario Project to highlight the implications for policy and service delivery, but he now presented them in even more direct terms. The moment for action had arrived.

Key Messages:
1. The new genetics represents the most significant scientific revolution in the history of medicine to be faced by government and the NHS.
2. Policy within the UK must encourage investment in science and be supportive of enterprise in the biotechnology and pharmaceutical industries.
3. We must immediately establish educational, training, and manpower strategies.
4. Regulatory frameworks must be light enough to encourage promising scientific developments but rigorous enough to protect the public.
5. Specific regulatory issues in relation to the confidentiality of genetic and personal health issues must be addressed.
6. Policies for service provision must develop services for people with inherited disorders as a first priority and start to prepare the NHS for the impact of genetic science across all aspects of the service.[67]

Milburn got the message. In part, this reflected the culmination of trends that had been unfolding for some time. Awareness of the importance of genetics and been rising and it had been moving up the policy agenda. Milburn appeared more interested in genetics than many of his predecessors and it fitted with current thinking about the need to invest in and modernise the health service. John Burn had also been serving as an advisor to Milburn and was 'able to get his ear' on genetics.[68]

> I was able to argue, coherently, I think, as did Ron and others, that the Human Genome Project was transformative. We were finding the gene for all these diseases, and pretty soon we were going to make a major impact on healthcare. Because of breast cancer and bowel cancer and all these things … people were realising we were breaking into areas of major significance.[69]

Burn was, and has long been, and influential voice because, as Milburn puts it, he 'understood politics without being political'.[70] There was also an implicit North East connection. Burn had his base in Newcastle – the centre of the Northern Genetics Service – and had helped to establish the new 'Centre for Life' in 2000. Milburn had been the MP for Darlington since 1992.

Securing the Future

Milburn's interest in genetics manifested itself in a set-piece speech at the Centre of Life in April 2001. The original plan had been for the speech to be delivered in London, but it was rescheduled as a result of the Foot and Mouth outbreak and

Burn persuaded Milburn to come to Newcastle.[71] This was a symbolic moment. It demonstrated that genetics had finally arrived in meaningful policy terms. Milburn spoke about the reticence of previous Health Secretaries to discuss genetics, couching it as the 'duty' of a 'responsible' government to now address the issue. Genetics was seen as 'A gift that modern science has bequeathed medicine and society'.[72] The Human Genome Project had 'crossed a new frontier in scientific knowledge'. The question, Milburn suggested, was 'whether we can harness that knowledge to also cross a new frontier in medicine':

> The implications of the advances in genetic knowledge are enormous – equal potentially for the conquest of disease to the discovery of antibiotics. This is revolution, with the potential in the first half of this century to dwarf the impact computer technology had on society in the second half of the last century'.[73]

It was clear that Milburn – and the civil servants inside the Department of Health who drafted the speech – had absorbed many of the messages that they had heard over the last few years. Naomi Brecker describes the policymaking process in these terms:

> The thinking would develop, you would call in the influencers in the field, the ideas would percolate, drafts would be shared, and then periodically you would have to seek buy-in from your lead Ministers, and then if it looks like it had legs ... civil servants would work on how best to make the announcements and what event would you hang the announcement on, and then put together the speech ... it is all very *Yes Minister* isn't it.[74]

In his speech, Milburn described how a better understanding of the genetic contribution to common complex conditions would eventually allow personally tailored interventions and new forms of prevention. Public understanding and support would be important if the benefits of genetics were to be realised, he argued. And it would be essential to prepare the NHS for the changes that lay ahead. With this in mind, the headline announcement in the speech was £30 million of new investment in genetic services. There would be more consultant geneticists, more genetic counsellors, and more laboratory staff, Milburn promised, as well as two new national reference laboratories.

The speech also provided more detail about the establishment of a series of 'Genetics Knowledge Parks' (GKPs) around the country. Such an idea had been foreshadowed in 2000 as part of the new 'NHS Plan', though seemingly without all that much thought being given to their role. According to an anonymous member of the Advisory Group on Genetics Research (AGGR), the body established to monitor the performance of the GKPs:

> It appeared very late in the drafting of the NHS plan, virtually just a sentence, just a throw away sentence that took everyone by surprise and when Alan Millburn was questioned what was it (?), he said, 'You tell me'. We then began to develop some themes.[75]

By the time of Milburn's 2001 speech the Genetics Knowledge Parks were conceived as 'centres of clinical and scientific excellence seeking to improve the diagnosis, treatment, and counselling of patients'.[76] Milburn also flagged an initiative that would become UK Biobank and set out plans for a Green Paper – a means of facilitating debate, thinking through different options, and seeking consultation about the approach the government should take. In time this would become a White Paper which set out the government's preferred options. The intention was to make sure that the potential benefits of genetics were realised, whilst ensuring that public opinion was sufficiently comfortable with what this meant, particularly at a time when science was still 'in the dock'.[77] It was clear therefore that the development of genetics policy in Britain had entered a new phase. Milburn's 2001 speech demonstrated that there would now be more substantial interest in the subject from the centre and that a number of different initiatives and mechanisms of support would be put in place to try and ensure that the potential of genetics was realised.

It is important to recognise that this did not necessarily mean that genetics itself had become a major policy issue. From a political perspective, there were more pressing concerns. Milburn's immediate focus was on reducing waiting times, steering the NHS through a series of winter crises, and getting more investment into the service.[78] In January 2000, the Prime Minister, Tony Blair, had announced that health spending would be significantly increased and brought into line with the European Union average. The financial pressure on the NHS had long been an important political issue. The future needs of the service were then spelled out in the 2002 Wanless Review.[79] Alongside increased funding, however, reform and modernisation were also needed, it was felt. This strategy was encapsulated in the 2000 NHS Plan, which suggested that better services could be driven in part by setting central targets and performance management, as well as by ensuring flexibility and elements of patient choice. As Milburn wrote in his Introduction to the Plan, 'At its heart the problem for today's NHS is that it is not sufficiently designed around the convenience and concerns of the patient'.[80]

Within this grand scheme, genetics was a 'small side issue'.[81] But it was an issue which did not contradict the wider narrative and could in fact complement it. Through its reforming agenda, the Labour government was 'relentlessly focussing' on the future, and genetics was an area of health policy which would naturally reach full salience in the medium to long term.[82] It was an issue which could help to frame ongoing political debates about belief in the NHS and what it represented. The benefits of genetics would need to be felt fairly and equitably – something which only the universal and comprehensive nature of the NHS could deliver – at a time when the Conservative Party was perceived to be focussing on tax incentives around private medicine instead.[83]

In a sense, the establishment of the Genetics Knowledge Parks was an 'act of faith' by Milburn.[84] An important role was played by Sir John Pattison, who recognised the need to broaden the scope of public investment into genetics.[85] From a Research and Development perspective there was interest in the potential of new therapies and new

diagnostics, and the knowledge parks were expected to help achieve a better align-
ment between research, industry, and the health service. There was also a narrative
around the need to maintain national economic competitiveness, and an expectation
that after the initial investment they would become self-sustaining, by driving fruitful
interactions between universities, industry, and other interests – as was understood
to have happened at other science parks around the country – as part of an entrepreneurial 'spin-out culture'.[86] The Genetics Knowledge Parks therefore represented a 'meeting point for many political and economic aspirations'.[87]

The initial expectation had been that four parks would emerge from a Genetics
Knowledge Challenge Fund of £10 million. The budget was increased to
£15 million with extra funding from the Department of Trade and Industry. A
call for tenders was released in August 2001 with just a few weeks' notice ahead of
an October deadline. Robertson's study of the parks found that 'Several AGGR
members commented that the speed of the commissioning process could be traced
back to political pressures at the time 'to do something in genetics''.[88] Peter
Harper felt that the Department of Health initially had quite an England-centric
view of the project.[89] Successful bids eventually came in from Newcastle, London,
Oxford, Cambridge, the North West, and Cardiff.

Six Genetics Knowledge Parks were therefore launched in November 2002.
Each had its own approach and strengths, based on its local characteristics and
collaborations between universities, research centres, health services, and biotechnology companies. For example, the Cambridge Genetics Knowledge Park
was couched as a partnership between the University of Cambridge and
Addenbrookes Hospital. The founding centres were the Public Health Genetics
Unit and the Centre for Medical Genetics and Policy, which had been formed in
2001 as part of the University's Faculty of Medicine and sought to promote research and teaching in medical genetics and provide an interdisciplinary forum for
debate. The overall picture was in fact much more complex. According to the
anthropologist Marilyn Strathern:

> You could never add up all the elements of the CKGP. Expertise is found
> lodged in bodies of diverse kinds – a veritable [twenty-first century]
> encyclopaedia of cross-referencing entities. These bodies are named
> variously as *faculties, departments, research centres, research groups, research
> programmes, units, institutes, schools, laboratories,* across some 17 *disciplines*
> and areas of *expertise,* including another university *campus,* while outreach to
> industry brings in other entities, such as an *enterprise,* a regional *initiative,* and
> a transatlantic *company;* there are in addition named *participants, partners,* and
> *sponsors,* and more diffusely *consumers* and *the public.*[90]

The aims of the Cambridge Genetics Knowledge Park centred on the translation of
genetic research and the realisation of commercial opportunities, understanding
patient and public perspectives, and placing them in their ethical, legal, and social
context, and contributing to the development of genetics policy. Zimmern became

its founding Director. He successfully overcome the objections of senior scientists in Cambridge who instinctively felt that any new funding should go into applied research. His position was also unique in that the Knowledge Park was his main focus, and his background was that of a public health physician. The Directors of the other five GKPs were leading geneticists who took on the role alongside their other responsibilities. Burn was appointed as Director of the Northern Genetics Knowledge Park (in Newcastle). The Director of the Wales Gene Park in Cardiff was Professor Julian Sampson. The Oxford Genetics Knowledge Park was led by Dr Jenny Taylor. Dian Donnai was Executive Director of the North West Genetics Knowledge Park, based in Manchester. The London IDEAS Park was led by Professor Steve Humphries. The six Directors met regularly and were expected to build a national network. This proved difficult in practice because their areas of immediate interest were different.[91] Nonetheless, it was felt to be an 'exciting time' with the Parks taking a place in the 'genomic revolution'.[92]

The annual funding of £1 million for the Cambridge Genetics Knowledge Park represented a significant change of scale for the PHGU. It allowed them to 'put flesh on the bones' and expand on the multidisciplinary work that they had been doing.[93] The staff increased from five to more than twenty. Teams focussed on genetic epidemiology, ELSI issues, public health, knowledge and dissemination, and business and administration. The ELSI team included newly appointed Lecturers in Law, Social Science, Philosophy, and Health Economics, managed by the Centre for Medical Genetics and Policy. This approach was unique to the Cambridge centre, though ensuring cohesion across the different teams and the many different perspectives proved to be difficult. Even so, Hilary Burton reflects that 'It did enable us to learn quite a bit, and to really properly embed those other disciplines into the work'.[94]

FIGURE 3.2 Dr Hilary Burton

Source: Reproduced with the permission of the PHG Foundation.

The PHGU's most significant research during this period came in relation to the evaluation of genetic tests. This was arguably the area in which a public health perspective had the most direct impact on genetics practice.[95] Recognising that tests were increasingly becoming available for common diseases as well as single-gene disorders, a number of observers began to emphasise the need to evaluate their reliability and their clinical benefits ahead of their widespread use.[96] There was an economic imperative – the need to ensure that scarce health resources, particularly in the context of the NHS, were well spent – but also a need to be wary of commercial pressures which were likely to encourage the wider use of tests that had not been properly evaluated.[97] As Zimmern and colleagues re-cognised in 2004, 'Evaluation may be imperfect and results incomplete, but failure to perform adequate assessments of new tests will reduce the quality of health care and have a detrimental effect on the public health'.[98]

Nonetheless, these competing pressures meant that achieving sufficient pro-fessional and public consensus and developing appropriate guidance was preferable to introducing formal legislation. As a result, the focus was often on the use of the ACCE framework, first developed in the United States during the early 2000s, which encompassed analytical validity, clinical validity, clinical utility, and ethical, legal, and social implications in evaluating genetic tests.[99] The PHGU drew on the ACCE framework as part of its review of genetics tests for Familial Hypercholesterolaemia. The Unit was instrumental in the wider adoption of the framework in Britain and also contributed to its further refinement. In the report *Moving Beyond ACCE: An Expanded Framework for Genetic Test Evaluation,* Wylie Burke, visiting the PHGU from the University of Washington, and Ron Zimmern argued for a broader understanding of the concept of clinical validity and highlighted the importance of defining the purpose of a test.[100]

Another source of influence came with the adoption of the Human Tissue Act in 2004. After high-profile scandals around organ retention at Alder Hey Children's Hospital in Liverpool and Bristol Royal Infirmary during the 1990s, the Act established the Human Tissue Authority to regulate the use and storage of such material. Though not the explicit intent, this had an impact on research at a cellular level being conducted by molecular geneticists and others. The concept of 'appropriate consent' outlined in the original Human Tissue Bill in 2003 was felt by researchers to be inappropriate for these kinds of fields, potentially constraining important medical research, and there were fears about the bureaucratic burden.

Representatives of the PHGU and Cambridge Genetics Knowledge Park tried to bring a professional practice perspective and were among those in the bio-medical and genetics communities that argued for a number of clarifications in the Bill.[101] A workshop in early 2004 brought together an interdisciplinary group of experts to debate the merits of various amendments to the Bill, a number of which were presented for consideration.[102] After deliberation with a range of expert groups the government made a number of changes. Kathleen Liddell, the new Lecturer in Law, and Alison Hall, a Research Associate at the Cambridge Genetics Knowledge Park with experience as a nurse and a solicitor and an interest in patient care, described how 'with tremendous persistence, the arguments of the research community began slowly to percolate through'.[103] Stewart and Zimmern suggested that the PHGU was 'able to help influence the political process to achieve amendments that at least partly resolved these problems'.[104]

A further important contribution came through debates about genetics education for health professionals. In 2002 the PHGU was commissioned by the Department of Health and the Wellcome Trust to put together plans for a national strategy. A team led by Hilary Burton produced the report *Addressing Genetics, Delivering Health* in 2003.[105] The need to recognise the growing importance of genetics in health care – moving from a small speciality to mainstream medicine – and the need to close the gap between the science and the understanding of health professionals was now well understood. This work, like the Nuffield Trust funded Genetics Scenario Project, drew on a series of workshops involving a range of professional groups and patients to identify needs and resources. *Addressing Genetics, Delivering Health* also called for the establishment of a new Centre for Genetics Education to coordinate a formal Programme for Genetics Education, and identified an important dichotomy between current and future clinicians. Whereas the former would initially need to understand how their current practice would evolve in the short term, the latter would need a much broader perspective.

Other significant reports produced by representatives of the Cambridge Genetics Knowledge Park and its partners during this period considered genetic diagnosis of learning disabilities, the ethical dimensions of stem cell medicine, and pharmacogenetics.[106] Burton reflects that she is most proud of work which considered the place of genetics in mainstream medicine, seeking to identify what the role of genetics would be and conducting reviews of the services that were currently provided – sometimes identifying significant inequities – in ophthalmology, inherited metabolic diseases, and cardiovascular disease.[107] The PHGU also had well-established links with the Joint Committee of Medical Genetics. The Joint Committee commissioned research on metabolic disease in 2004 and, in 2006, formed a working group to consider the implications of the Human Tissue Act now that it was in action. Alison Hall was the lead author of the resulting report. The Joint Committee also worked with the PHGU on the use of cell-free foetal nucleic acids for the purpose of non-invasive prenatal diagnosis. As part of this, the PHGU led a working group of patients, physicians, laboratories, and commissioners, producing policy recommendations for the Joint Committee and the Department of Health in 2009.[108]

A common theme running through this research was the need to engage with professionals, partner organisations, and groups of stakeholders, to seek consensus and find practical solutions to policy questions. Many of the approaches inherent to public health genetics were becoming well established in debates about genetics policy in Britain, and in relation to a broader range of issues. The PHGU had increasingly close relationships with health policymakers, including inside the Department of Health, the bodies and committees that formed the emerging infrastructure of policy support around genetics, and with other influential organisations.

Perhaps the clearest demonstration of this change, and the most significant part of the new policy infrastructure was the establishment of the UK Genetic Testing Network (UKGTN) in 2002. The Network's Steering Committee was formed as sub-group of the Genetics Commissioning Advisory Group, and comprised representatives of laboratories, geneticists, commissioners, and patients. Its aim was to ensure high quality and equity in the provision of genetic tests across the NHS – 'to ensure that patients can get tests on the basis of clinical need, not where they live'.[109] Peter Farndon became the first Chair. A Project Team with a Director, Scientific expert, Clinical expert, Public Health expert, and Project Manager, conducted the detailed research, subsequently aided by the creation of five working groups which formed the 'engine room' of the UKGTN: Gene Dossier and Directory, Laboratory Membership and Audit, Communications and Systems, Service Development and Clinical Appropriateness, and Commissioning. Zimmern became a member of the Steering Committee and Chair of the Service Development and Clinical Appropriateness working group. UKGTN was directly accountable to the Department of Health, plugged into formal specialist commissioning processes, and had established relationships with other advisory bodies and expert groups.

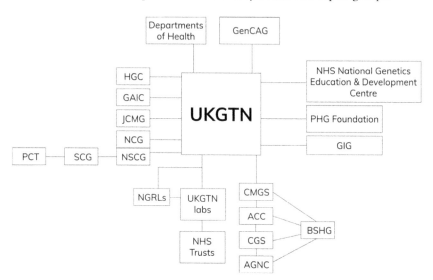

FIGURE 3.3   The Place of the UK Genetic Testing Network – Adapted from *First Report of the UKGTN: Supporting Genetic Testing in the NHS* (UK Genetic Testing Network, 2007). ACC, Association of Clinical Cytogeneticists; AGNC, Association of Genetic Nurse Counsellors; BSHG, British Society of Human Genetics; CGG, Cancer Genetics Group; CMGS, Clinical Molecular Genetics Society; GAIC, Genetics and Insurance Committee; GenCAG, Genetics Commissioning Advisory Group; GIG, Genetic Interests Group; HGC, Human Genetics Commission; JCMG, Joint Committee for Medical Genetics; NCG, National Commissioning Group; NGRL, National Genetics Reference Laboratory; NSCG, National Specialised Commissioning Group; PCT, Primary Care Trust; SCG, Specialised Comissiong Group

New genetic tests – initially for single-gene disorders – were evaluated as they became available and approved for clinical use by one of the Network's member laboratories. From 2003, these tests were collated in an annual NHS Directory of Molecular Genetics Testing. The evaluation process was based on the ACCE framework. Each test required a complete 'gene dossier' which included its clinical utility and validity. The process was conceived as an independent assessment of tests, separate from considerations of their funding.[110] The PHGU played an important role in the development and application of the gene dossier approach, having introduced the ACCE framework and identified subsequent modifications which took better account of the target population, disease, and purpose of testing. Unsurprisingly links were strongest through the UKGTN's Service Development and Clinical Appropriateness working group, which aimed to keep abreast of developing relationships between genetic testing services and NHS services and consider potential changes in response to new practices and new technologies.

The role of the UKGTN subsequently expanded to include evaluation of genomic tests and complex molecular biomarkers, in addition to tests for single-gene disorders.[111] Following the introduction of panel tests and then whole genome sequencing, there were fewer applications for new tests, but each test could achieve more. Burton and the PHGU led a working group for UKGTN which evaluated the use of array CGH – Comparative Genomic Hybridisation – in routine clinical medicine, and similar research on Ophthalmology was also commissioned.[112] Mark Kroese, a public health physician who was appointed as the UKGTN's 'Public Health expert' in 2006 had close links with the PHGU. He recalls that it was a 'mutually beneficial relationship for both organisations'.[113] Kroese had previously undertaken a placement at the PHGU as a public health trainee and worked with Zimmern and Burton, learning about genetics in the context of population health and gaining experience in commissioning. He had also spent time in the United States with both Muin Khoury and Wylie Burke. Kroese looks back on this period during the mid-2000s as one of opportunity in which, initially from quite a specialised perspective but then more widely, genetics came to have a more important place in health service development. As he reflects, 'As a young public health physician this all looked very interesting and exciting'.[114] It was also a period in which public health, in contrast to much of its recent history, had more adequate resources at its disposal and an influential role in identifying health needs and developing services, something which was then lost in the course of subsequent NHS reforms. The responsibility for public health was moved back to local authorities, outside of wider health service processes in 2012. Kroese joined the PHG Foundation full time in 2012, before becoming Director in 2017.

A number of reflections on the UKGTN suggest that it owed much to the hard work of Farndon and his colleagues in developing a collective approach, and to geneticists who recognised the need for agreed standards and policies at a time when the number of tests was increasing. There is also a perception that the UKGTN was successful because it was based on consensus and able to draw on the tradition of cooperation in the genetics community. It built on the existing working arrangements between genetics centres and laboratories around the provision of genetic tests for rare disorders – coordination and swapping of samples depending on local need – and put them on a more formal footing. Jacquie Westwood, UKGTN Project Director, also worked for the London Specialised Commissioning Group and had earlier been lead commissioner for genetics at South East Thames Regional Health Authority. At a time when funding levels and commissioning arrangements varied greatly across the country, she was known to have taken a real interest in genetics and developed more joined-up thinking. Westwood had also attended the Genetics and Health Policy course organised by the PHGU. One of the providers from whom she commissioned services was Frances Flinter, who also later became a member of the Commissioning Advisory Group of the UKGTN.[115] According to Farndon, investing 'a small amount of money into a piece of administration would … really pay dividends in having a mechanism for working together'.[116] In doing so, the UKGTN was able to offset some of the lack of coordination resulting from the operation of the NHS Internal Market over the intervening ten years – what Brecker describes as 'hang-ups over funding and patients' – and lay foundations for future developments.[117]

Many of the issues that had been discussed throughout the 1990s and early 2000s – the need for investment in laboratory services, genetics education, and research in order to prepare for coming changes in healthcare based around a new understanding of illness, prevention, diagnosis, and treatment, with attendant moral issues and the NHS as the necessary vehicle of change – made their way into the government's White Paper *Our Inheritance, Our Future: Realising the Potential of Genetics in the NHS*, published in June 2003.[118] There was a natural transition from ongoing policy work inside the Department of Health to the production of a White Paper. More 'mainstream' civil servants were also brought in to lead the initiative.[119] An Advisory Panel was established that included Bobrow, Burn, Farndon, O'Neill and Zimmern, alongside physicians and representatives of the biomedical and pharmaceutical industries. John Reid had replaced Alan Milburn as Secretary of State for Health just a few days before publication. While Reid had less experience in health policy and less immediate interest in genetics, the direction of travel was now set. According to Lord Warner, who had also only recently been appointed as a Health Minister, 'we had read it and kind of understood it, but we totally lacked any background in it'.[120] Burn recalls of Reid's approach at the official launch of the White Paper, 'He was just like a kid who'd got a toy for Christmas'.[121] The publication of a White Paper solely devoted to genetics demonstrated the extent to which it had become a meaningful area of policy with political traction. Brecker describes this as a 'mammoth achievement'.[122] In his Foreword the Prime Minister, Tony Blair, made a personal commitment to the vision which was set out, identifying the need to invest and prepare for 'the coming revolution in health care' and framing it in the context of Britain's 'remarkable scientific tradition':

The more we understand about the human genome, the greater will be the impact on our lives and on our healthcare. As an increasing number of diseases are linked to particular genes or gene sequences, we will be able to target and tailor treatment better to offset their impact and even to avoid the onset of ill-health many years in advance ... I am also absolutely determined that the National Health Service should be able to respond to these advances so the benefits of genetic and the more personalised and improved healthcare it will bring are available to all.[123]

At the White Paper's launch a further £50 million was announced to expand the specialist genetics workforce, modernise laboratories and IT systems, further develop clinical training, and support a range initiatives which would develop the place of genetics in everyday medicine and primary care. This was understood in terms of 'mainstreaming' genetics inside the NHS.[124] It brought real benefits for geneticists. According to Flinter, 'we had to move into new premises because we suddenly had more staff and more labs, and we needed more room'.[125] It also resonated with ambitions of keeping Britain at the cutting edge of research and development and reducing waiting times for patients.

The Human Genetics Commission and the National Screening Committee were asked to feed into discussions about genetics and insurance and the potential for genetic profiling at birth. The latter was an issue which spoke to ongoing differences of emphasis between geneticists – some of whom saw sequencing the genome of every newborn baby as a potentially positive step – and others from a more concerted public health perspective who could foresee the ethical and social questions which would arise.[126] The White Paper's establishment of a national strategy for genetics education, promoted by the PHGU, was described as a 'catalyst to bring education and training in genetics for all NHS healthcare staff'.[127] After a competitive bidding process the National Genetics Education and Development Centre was established in Birmingham in 2004, led by Farndon and colleagues. An initial scoping exercise found that some clinicians felt that genetics was still of little practical application. This view was often strongest amongst senior members of the professions who had no recent experience of the ways in which genetics was already changing practice.[128] Each professional group needed a tailored approach.[129] Medical students would require 'just in case' information, established practitioners would require 'just in time' information'.[130] As with the UKGTN, the National Genetics Education and Development Centre continued its work developing and coordinating genetics education for professionals outside genetic specialities, before expanding to include genomic medicine.

The 2003 White Paper also gave further impetus to the development of UK Biobank, a project which had been set in train by the Medical Research Council and the Wellcome Trust and sought to facilitate research which would lead to better understanding, at a population level, of the interaction between genetic, environmental and lifestyles factors in the development of common diseases. There was a growing understanding that large cohort studies that allowed many participants to be followed over a long period of time – including those who did not develop diseases as well as those who did – would be increasingly important in developing better prevention, diagnosis, and treatment. There was also a long tradition of these kinds of cohort studies in wider British public health.[131] John Burn, for example, recalls 'banging the table saying 'Why aren't we doing hereditary familial hypercholesterolemia? Why aren't we sorting that out''.[132] At the same time, many geneticists were concerned that high-profile initiatives like UK Biobank would be expensive, potentially taking funding away from other important research, and that the exercise was likely only to identify small or moderate genetic risk factors. There were also concerns about privacy and the use of genetic information, while a number of scientists questioned the overall design of the study. An article in the *Lancet* described UK Biobank as a 'project in search of a protocol'.[133]

John Bell, Regius Professor of Medicine at the University of Oxford from 2002, was an important supporter of UK Biobank and became a member of its Board. He suggests that the project owed much to the initiative of Sir George Radda, Chief Executive of the MRC between 1996 and 2004, in pushing it through despite its failure to secure funding after being discussed by at least two MRC committees. This approach, which was driven from the top-down, circumventing the usual means of securing funding for large scale research projects, would be returned to again in the early 2010s with the development of the 100,000 Genomes Project. Bell reflects that the 'UK has done extraordinarily well in this space and almost everything we did would have been hard to do through conventional peer review funding. It was mostly strategic funding'.[134] We can see therefore that the rise of genomics owes much to its political and scientific tractability, as well as its clinical utility.

John Newton, who became the first Chief Executive of UK Biobank in 2003, suggests that after a detailed consultation and planning process a number of critics had been won round. [135] Even so, he says that 'People like me and John Bell didn't pay too much attention to what other people thought, we just got on with it'.[136] Newton was a trained public health physician and had previously been a Director of Research at the Radcliffe Hospital in Oxford, where he became familiar with the work being done by Bell and colleagues at the Wellcome Centre for Human Genetics.

The ethical underpinnings of UK Biobank were influenced in part by the work of Bartha Knoppers around the meaning of individual consent for such longitudinal studies.[137] Eric Meslin, who was Vice-Chair of the Ethics and Governance Council of UK Biobank between 2015 and 2018, has also noted the contrasts between traditional clinical trials and biobank studies. Even though the right to withdraw exists in a biobank study, the ability of a participant to leave once a sample has been analysed and the information has been made available becomes more complex.[138] As he succinctly puts it; 'It kind of reminds me of the line from the Eagles' *Hotel California* … 'You can check out any time you like, but you can never leave''.[139]

From 2006, blood, urine and saliva samples began to be collected from 500,000 participants aged between 45 and 69, alongside other health and lifestyle data. UK Biobank became open to researchers in 2012, allowing valuable data around the background rate of genetic markers in a normal population to be interpreted and the genetic contribution to a wide range of common conditions including heart disease and diabetes to be better understood. In 2019, it was announced that whole genome sequences would be assembled for each of the 500,000 volunteers. In 2021, the data of the first 200,000 was made available. Bell's reflection on UK Biobank is that:

> It is one of the things I'm most proud about … because you know that was really the first genomic epidemiology cohort and I set that up with a load of epidemiologists who couldn't care less about genomics. But in the end of course their payday has been all genomic related, so that was another piece of the puzzle.[140]

The early-to-mid 2000s was a time of significant change for genetics – in political and health service terms. The Department of Health now had a much greater interest in the field, and was much more involved in the networks of support and professional infrastructure. As John Burn reflects, 'I think it's fair to say that, when the 2003 White Paper came out, the geneticists were very much in the ascendency'. [141] The central tenets of the Government's approach and the evolving picture of genetics policy, and practice inside the NHS, now broadly reflected the interests, approaches, and capabilities of the mainstream clinical genetics community. The PHGU and other groups had a clear role in shaping these developments. Many of the principles inherent to public health genetics had come to have a significant influence.

A remaining key question, however, was how the NHS would adapt to accommodate the science of genomics as it evolved. What would happen as new technologies developed, and the once largely theoretical possibilities of genomic medicine started to become a more tangible reality? In his reflections on the White Paper *Our Inheritance, Our Future*, Lord Warner suggests that despite the clear articulation of the wonderful new science at its heart, there was still something of a disconnect between the rhetoric and the implementation: 'It was a bit elusive about how that was going to happen. And I think it stayed elusive for quite a long period of time'.[142] How would genetics be truly integrated into mainstream medicine, as everyone recognised that it eventually would be?. Who would be called upon to shape developments in future?

These questions eventually crystallised in large part because, throughout the 1990s and early 2000s while many of these policy developments were taking place, the Human Genome Project was advancing. Though its implications were still largely scientific, and it would be some time before they came to directly influence clinical medicine, the Human Genome Project was often at the back of the minds of civil servants and practitioners as they considered the future direction of genetics.

A first working draft of the human genome was completed in June 2000. US President Bill Clinton and the British Prime Minister Tony Blair announced the milestone at a White House press conference. Clinton asserted that 'With this profound new knowledge, humankind is on the verge of gaining immense, new power to heal'.[143] Evelyn Fox Keller wrote that 'It would be hard to imagine a more dramatic climax to the efforts of the entire century'.[144]

The official completion of the Human Genome Project in 2003 was similarly heralded as a great moment. John Sulston, Director of the Sanger Centre and one of the leading lights of the Project in Britain, and later Chair of the HGC between 2007 and 2009, suggested that it 'defines a moment of the history of life on earth'.[145] Peter Harper later reflected that it was arguably 'the most important event in biology since modern science began'.[146] David Bentley, one of Sulston's colleagues, described what it meant in practice:

> The ability to predict problems, which are intrinsic in our genetic makeup earlier. To enable cures to be applied when they really need to be applied, or treatments at least to be applied when they can be applied early enough to really be effective or even preventative.[147]

How, when, and in what ways this would happen was yet to be decided. Many such assertions were clearly bound up with wider political and strategic research interests, and, for some at least, it was clear that the completion of the project would not lead to an overnight transformation. Martin Bobrow − who had sounded a note of caution in 1995 about thinking too far ahead − offers an important perspective:

> The impact of the Human Genome Project? That's either a sentence or the rest of the day. So that turned it upside down, but it didn't turn it upside down overnight, it turned it upside down over a period of five years.[148]

Similarly, Wylie Burke identifies a long and incremental period of change after 2003 − a 'constant drumbeat' − of more functions being ascribed to specific genes and increasing opportunities to use genomics in a clinical context, particularly through new diagnostic capabilities and advances in areas such as pharmacogenetics.[149] The Human Genome Project was a key driver of these changes. But it did not, in itself, resolve the enduring questions around evaluation and clinical application, the organisation and funding of services, and of research. The networks and policy frameworks that had developed in Britain would have to adapt and keep pace. Zimmern reflects on the influence of the Human Genome Project in these terms: 'both from a technological and from a clinical perspective, it has changed the whole of biology, it is basically as influential as Darwin's theory of evolution'.[150] The reference to Darwin is apt however as over the course of the next few years, change would, for the most, part still be evolutionary rather than revolutionary.

Notes

1 Interview with Dr Ron Zimmern, February 2021.
2 Interview with Professor Sir Peter Harper, November 2020.
3 Interview with Dr Layla Jader, June 2021. https://drlaylajader.co.uk/
4 Interview with Professor Sir John Burn, December 2020.
5 Interview with Professor Frances Flinter, August 2021.
6 *The Development and Influence of Public Health Genomics* (University of Liverpool, 2022).
7 Interview with Professor Sir John Burn, December 2020.
8 *Stem Cell Research: Medical Progress with Responsibility* (Department of Health, 2000).
9 Hansard. HC Deb. December 19, 2000. Vol. 63. cc.211-66. S. Louët, 'UK Set to Allow Research on Embryonic Cells', *Nature Biotechnology*, Vol. 18, No. 1034, 2000.
10 J. Lenaghan, *Brave New NHS? The Impact of the New Genetics on the Health Service* (IPPR, 1998).
11 Ibid., p.ii.
12 Ibid.
13 Ibid., p.46. M.J. Khoury, 'From Genes to Public Health: The Application of Genetic Technology in Disease Prevention', *American Journal of Public Health*, Vol. 86, No. 12, 1996, pp. 1717–22.
14 *Clinical Genetics Service into the 21st Century* (Royal College of Physicians, 1996).
15 *Clinical Genetics Service: Activity, Outcome Effectiveness and Quality* (Royal College of Physicians, 1998).
16 *Commissioning Clinical Genetic Services* (Royal College of Physicians, 1998).
17 *Genetics and Cancer Services: Report of a Working Group for the Chief Medical Officer* (Department of Health, 1998). D. Wonderling, P. Hopwood, A. Cull, F. Douglas, M. Watson, J. Burn, and K. McPherson, 'A Descriptive Study of UK Cancer Genetics Services: An Emerging Clinical Response to the New Genetics', *British Journal of Cancer*, Vol. 85, No. 2, 2001, pp.166–70.
18 *Commissioning Clinical Genetic Services* (Royal College of Physicians, 1998).
19 *Genetics in Primary Care* (Royal College of General Practitioners, 1998) p. iii.
20 *Ibid.*, p. 11.
21 *Genetics and Health: Policy Issues for Genetics Science and Their Implications for Health and Health Service* (Nuffield Trust, 2000) p. xi.
22 Interview with Dr Ron Zimmern, November 2020.
23 *Genetics and Health: Policy Issues for Genetics Science and Their Implications for Health and Health Service* (Nuffield Trust, 2000) p. 2.
24 Ibid., p. 9.
25 Ibid.
26 Ibid., p. Ix.
27 Ibid., p. 1.
28 Ibid., p. 75.
29 Ibid.
30 Ibid., p. 76.
31 Ibid.
32 Ibid., p. 77.
33 Ibid., p. 4.
34 Interview with Dr Ron Zimmern, November 2020.
35 Interview with Carol Lyon, April 2021.
36 *Beyond the Horizon: Connecting Science and Health* (PHG Foundation, 2012) p. 27.
37 Interview with Dr Ron Zimmern, November 2020.
38 Ibid.
39 Interview with Dr Sheila Adam, February 2021. P.S. Harper, *The Evolution of Medical Genetics: British Perspective* (CRC Press, 2020).
40 Interview with Naomi Brecker, February 2020.
41 Ibid.

42 Ibid.
43 Interview with Dr Mark Bale, February 2020.
44 Interview with Dr Ron Zimmern, February 2020.
45 Interview with Dr Mark Bale, February 2020.
46 *Laboratory Services for Genetics: Report of an Expert Working Group to the NHS Executive and the Human Genetics Commission* (Department of Health, 2000). D. Donnai and R. Elles, 'Integrated, Regional Genetics Services: Current and Future Provision', *British Medical Journal*, Vol. 322, April 28, 2001. D. Ravine and J. Sampson, 'Commentary: The Future Development of Regional Genetics Services will rely on Partnerships', *British Medical Journal*, Vol. 322, 28 April 2001.
47 Interview with Dianne Kennard, April 2021.
48 Interview with Naomi Brecker, February 2021.
49 Harper, *The Evolution*, p. 195. A. Stewart, P. Brice, H. Burton, P. Pharoah, S. Sanderson, and R. Zimmern, *Genetics, Health Care and Public Policy: An Introduction to Public Health Genetics* (Cambridge University Press, 2006).
50 *The Development and Influence of Public Health Genomics* (University of Liverpool, 2022).
51 *The Advisory and Regulatory Framework for Biotechnology: Report from the Government's Review* (Cabinet Office, 1999).
52 S. Abdulla, 'UK Government Demonstrates Committee-Mania', *Nature Medicine*, Vol. 5, No. 856, 1999, p. 856.
53 Interview with Baroness O'Neill, January 2021.
54 Interview with Professor Sir Peter Harper, November 2020
55 Harper, *The Evolution*
56 Interview with Alastair Kent, November 2020.
57 *Debating the Ethical Future of Human Genetics: First Annual Report of the Human Genetics Commission -2001* (Human Genetics Commission, 2001).
58 *Outcome of the Public Consultation on Preimplantation Genetics Diagnosis* (London: Human Fertilsation and Embryology Authority, 2001).
59 Interview with Professor Sir John Burn, December 2020.
60 *Genes Direct: Ensuring the Effective Oversight of Genetics Tests Supplied Directly to the Public – A Report by the Human Genetics Commission* (Department of Health, 2003). *Nothing to Hide, Nothing to Fear? Balancing Individual Rights and the Public Interest in the Governance and use of the National DNA Database – A Report by the Human Genetics Commission* (Department of Health, 2009).
61 Interview with Professor Sir Peter Harper, November 2020.
62 Interview with Professor Sir Bruce Ponder, December 2020.
63 Interview with Dr Mark Bale, February 2021.
64 Ibid.
65 Interview with Professor Sir Bruce Ponder, December 2020.
66 Interview with Dr Ron Zimmern, November 2020.
67 'What can we Expect from Genetics?', Presentation by Ron Zimmern to Secretary of State's Seminar on Genetics, September 13, 2000.
68 Interview with Professor Sir John Burn, December 2020.
69 ?Ibid.
70 Interview with Alan Milburn, July 2021.
71 Interview with Professor Sir John Burn, December 2020.
72 https://www.ukpol.co.uk/alan-milburn-2001-speech-at-the-institute-of-human-genetics/
73 Ibid.
74 Interview with Naomi Brecker, February 2021.
75 M. Robertson, 'Translating Breakthroughs in Genetics into Biomedical Innovation: The Case of the UK Genetic Knowledge Parks', *Technology Analysis and Strategic Management*, Vol. 19, No. 2, 2007, p. 195.

76 https://www.ukpol.co.uk/alan-milburn-2001-speech-at-the-institute-of-human-genetics/

77 Interview with Alan Milburn, July 2021.

78 Ibid.

79 *Securing Our Future Health: Taking a Long-Term View* (HM Treasury, 2002). N. Timmins, *The Most Expensive Breakfast in History: Revisiting the Wanless Review 20 Years On* (Health Foundation, 2021).

80 *The NHS Plan: A Plan for Investment, A Plan for Reform*, Cmnd. 4818-I (London: HSMO, 2000) p. 15.

81 Interview with Lord Warner, June 2021.

82 Interview with Alan Milburn, July 2021.

83 Ibid.

84 Interview with Alastair Kent, November 2020.

85 Interview with Professor Dame Sally Davies, February 2021.

86 Interview with Dr Mark Bale, February 2021.

87 E. Khlinovskaya Rockhill, 'On Interdisciplinarity and Models of Knowledge Production', *Social Analysis*, Vol. 51, No. 3, 2007, p. 121.

88 Robertson, 'Translating Breakthroughs', p. 195.

89 Interview with Professor Sir Peter Harper, November 2020.

90 M. Strathern, 'Experiments in Interdisciplinarity', *Social Anthropology*, Vol. 13, No. 1, 2005, p. 79.

91 Interview with Ron Zimmern, February 2021. Robertson, 'Translating Breakthroughs' i.

92 Interview with Dr Philippa Brice, April 2021.

93 Interview with Alison Stewart, February 2021.

94 Interview with Dr Hilary Burton, January 2021.

95 Interview with Alison Stewart, February 2021.

96 M. Kroese, R. Zimmern, and S. Sanderson, 'Genetic Tests and Their Evaluation: Can we Answer the key Questions', *Genetics in Medicine*, Vol. 6, No. 6, 2004, pp. 475–80.

97 W. Burke and R. Zimmern, 'Ensuring the Appropriate Use of Genetic Tests', *Nature Reviews Genetics*, Vol. 5, No.12, 2004, pp. 955–59.

98 Kroese, Zimmern, and Sanderson, 'Genetic Tests', pp. 479–80.

99 J. Haddow and G. Palomaki G, 'ACCE: A Model Process for Evaluating Data on Emerging Genetic Tests' in M. Khoury, J. Little, and W. Burke (eds.), *Human Genome Epidemiology* (Oxford University Press, 2004) pp. 217–33.

100 W. Burke and R. Zimmern, *Moving Beyond ACCE: An Expanded Framework for Genetic Test Evaluation* (PHG Foundation, 2007).

101 Interview with Alison Hall, May 2021.

102 B. Parry, ' The New Human Tissue Bill: Categorization and Definitional Issues and their Implications', *Genomics, Society and Policy*, Vol. 1, No. 1, 2005, pp. 74–85.

103 Interview with Alison Hall, May 2021. K. Liddell and A. Hall, 'Beyond Bristol and Alder Hey: The Future Regulation of Human Tissue', *Medical Law Review*, Vol. 13, No. 2, 2005, p. 199.

104 A. Stewart and R. Zimmern, 'What is Public Health Genomics?' in H.F. Willard and G.S. Ginsburg (eds.), *Genomic and Personalized Medicine: Volume 1 – Principles, Methodology and Translational Approaches* (Academic Press, 2009) p. 451.

105 H. Burton, *Addressing Genetics, Delivering Health: A Strategy for Advancing Dissemination and Application of Genetics Knowledge Throughout our Health Professions* (PHG Foundation, 2003).

106 B. Gogarty, *Parents as Partners: A Report and Guidelines on the Investigation of Children with Developmental Delay: By Parents, for Professionals* (Cambridge Genetics Knowledge Park, 2006). O. Corrigan, K. Liddell, J. McMillan, A. Stewart, and S. Wallace, *Ethical Legal and Social Issues in Stem Cell Research and Therapy: A Briefing Paper from Cambridge Genetics Knowledge Park* (Cambridge Genetics Knowledge Park, 2006). D. Melzer,

A. Raven, D. Detmer, T. Ling, and R. Zimmern, *My Very Own Medicine: What Must I Know? Information Policy for Pharmacogenetics* (University of Cambridge, 2003).

107 Interview with Dr Hilary Burton, January 2021. T. Moore and H. Burton, *Genetics Ophthalmology in Focus: A Needs Assessment and Review of Specialist Services for Genetic Eye Disorders* (PHG Foundation, 2008). H. Burton, *Metabolic Pathways: Networks of Care – A Needs Assessment and Review of Services for People with Inherited Metabolic Disease in the United Kingdom* (Public Health Genetics Unit, 2005). H. Burton, C. Alberg, and A. Stewart, *Heart to Heart: Inherited Cardiovascular Conditions Services – A Needs Assessment and Service Review* (PHG Foundation, 2009).

108 *Cell-Free Fetal Nucleic Acids for Non-Invasive Prenatal Diagnosis: Report of the UK Expert Working Group* (PHG Foundation, 2009).

109 *First Report of the UKGTN: Supporting Genetic Testing in the NHS* (UK Genetic Testing Network, 2007).

110 Peter Harper Interview with Professor Peter Farndon.

111 *First Report of the UKGTN: Supporting Genetic Testing in the NHS* (UK Genetic Testing Network, 2007).

112 *Evaluation of the use of Array Comparative Genomic Hybridisation in the Diagnosis of Learning Disability: Report of a UK Genetic Testing Network Working Party* (PHGU, 2006).

113 Interview with Dr Mark Kroese, April 2021.

114 Ibid.

115 Interview with Professor Frances Flinter, August 2021.

116 Peter Harper Interview with Professor Peter Farndon.

117 Interview with Namoi Brecker, February 2021. Harper, *The Evolution.*

118 *Our Inheritance, Our Future: Realising the Potential of Genetics in the NHS* (Department of Health, 2003).

119 Interview with Naomi Brecker, February 2021.

120 Interview with Lord Warner, June 2021.

121 *The Development and Influence of Public Health Genomics* (University of Liverpool, 2022).

122 Interview with Naomi Brecker, February 2021.

123 *Our Inheritance, Our Future: Realising the Potential of Genetics in the NHS* (Department of Health, 2003) p. 1.

124 Interview with Dianne Kennard, April 2021.

125 Interview with Professor Frances Flinter, August 2021.

126 Interview with Dr Rosalind Skinner, June 2021.

127 *Our Inheritance, Our Future: Realising the Potential of Genetics in the NHS* (Department of Health, 2003) p. 57.

128 Peter Harper Interview with Professor Peter Farndon.

129 P.A. Farndon and C. Bennet, 'Genetics Education for Health Professionals: Strategies and Outcomes from a National Initiative in the United Kingdom', *Journal of Genetic Counselling*, Vol. 17, No. 2, 2008, pp. 161–69.

130 Interview with Professor Peter Farndon, May 2021.

131 V. Berridge, M. Gorsky, and A. Mold, *Public Health in History* (Open University Press, 2011). A. Mold, P. Clark, G. Millward, and D. Payling, *Placing the Public in Public Health in Post-War Britain, 1948-2012* (Palgrave Macmillan, 2019).

132 Interview with Professor Sir John Burn, December 2020.

133 V. Barbour, 'UK Biobank: A Project in Search of a Protocol', *Lancet*, Vol. 361, May 17, 2003.

134 Interview with Professor Sir John Bell, August 2021.

135 W. Ollier, T. Srposen, and T. Peakman, 'UK Biobank: From Concept to Reality', *Pharmacogenomics*, Vol. 6, No. 6, 2005, pp. 639-646.

136 Interview with Professor John Newton, August 2021.

137 Interview with Professor John Newton, August 2021. S. Wallace, S. Lazor, and B.M. Knoppers, 'Consent and Population Genomics: The Creation of Generic Tools', *IRB: Ethics and Human Research*, Vol. 31, No. 2, 2009, pp. 15-20.

138 E.M. Meslin and I. Garba, 'Biobanking and Public Health: Is a Human Rights Approach the tie that Binds?', *Human Genetics*, Vol. 130, No. 3, 2011, pp. 451-463.
139 Interview with Dr Eric Meslin, January 2021.
140 Interview with Professor Sir John Bell, August 2021.
141 *The evelopment and Influence of Public Health Genomics* (University of Liverpool, 2022).
142 Interview with Lord Warner, June 2021.
143 'How Diplomacy Helped to end the Race to Sequence the Human Genome', *Nature*, June 24, 2020.
144 E.F. Keller, *The Century of the Gene* (Harvard University Press, 2000) p. 4.
145 Interview with Professor Sir John Salston, Cold Spring Harbor Laboratory Oral History Collection.
146 Harper, *The Evolution*, p. 283.
147 Interview with Dr David Bentley, Cold Spring Harbor Laboratory Oral History Collection.
148 Interview with Professor Martin Bobrow, November 2020.
149 Interview with Professor Wylie Burke, November 2020.
150 Interview with Dr Ron Zimmern, February 2021.

Bibliography

Abdulla, S., 'UK Government Demonstrates Committee-Mania', *Nature Medicine*, Vol. 5, No. 856, 1999.

Barbour, V., 'UK Biobank: A Project in Search of a Protocol', *Lancet*, Vol. 361, 2003.

Berridge, V., Gorsky, M., and Mold, A., *Public Health in History* (Open University Press, 2011).

Beyond the Horizon: Connecting Science and Health (PHG Foundation, 2012).

Burke, W. and Zimmern, R., 'Ensuring the Appropriate Use of Genetic Tests', *Nature Reviews Genetics*, Vol. 5, No. 12, 2004.

Burke, W. and Zimmern, R., *Moving Beyond ACCE: An Expanded Framework for Genetic Test Evaluation* (PHG Foundation, 2007).

Burton, H., *Addressing Genetics, Delivering Health: A Strategy for Advancing Dissemination and Application of Genetics Knowledge Throughout our Health Professions* (PHG Foundation, 2003).

Burton, H., *Metabolic Pathways: Networks of Care – A Needs Assessment and Review of Services for People with Inherited Metabolic Disease in the United Kingdom* (Public Health Genetics Unit, 2005).

Burton, H., Alberg, C., and Stewart, S., *Heart to Heart: Inherited Cardiovascular Conditions Services – A Needs Assessment and Service Review* (PHG Foundation, 2009).

Cell-Free Fetal Nucleic Acids for Non-Invasive Prenatal Diagnosis: Report of the UK Expert Working Group (PHG Foundation, 2009).

Clinical Genetics Service into the 21st Century (Royal College of Physicians, 1996).

Clinical Genetics Service: Activity, Outcome Effectiveness and Quality (Royal College of Physicians, 1998).

Commissioning Clinical Genetic Services (Royal College of Physicians, 1998).

Corrigan, O., Liddell, K., McMillan, J., Stewart, A., and Wallace, S., *Ethical Legal and Social Issues in Stem Cell Research and Therapy: A Briefing Paper from Cambridge Genetics Knowledge Park* (Cambridge Genetics Knowledge Park, 2006).

Debating the Ethical Future of Human Genetics: First Annual Report of the Human Genetics Commission - 2001 (Human Genetics Commission, 2001).

Donnai, D. and Elles, R., 'Integrated, Regional Genetics Services: Current and Future Provision', *British Medical Journal*, Vol. 322, 28 April 2001.

Evaluation of the use of Array Comparative Genomic Hybridisation in the Diagnosis of Learning Disability: Report of a UK Genetic Testing Network Working Party (Public Health Genetics Unit, 2006).

Farndon, P.A. and Bennet, C., 'Genetics Education for Health Professionals: Strategies and Outcomes from a National Initiative in the United Kingdom', *Journal of Genetic Counselling*, Vol. 17, No. 2, 2008.

First Report of the UKGTN: Supporting Genetic Testing in the NHS (UK Genetic Testing Network, 2007).

Genes Direct: Ensuring the Effective Oversight of Genetics Tests Supplied Directly to the Public – A Report by the Human Genetics Commission (Department of Health, 2003).

Genetics and Cancer Services: Report of a Working Group for the Chief Medical Officer (Department of Health, 1998).

Genetics and Health: Policy Issues for Genetics Science and Their Implications for Health and Health Service (Nuffield Trust, 2000).

Genetics in Primary Care (Royal College of General Practitioners, 1998).

Gogarty B., *Parents as Partners: A Report and Guidelines on the Investigation of Children with Developmental Delay: By Parents, for Professionals* (Cambridge Genetics Knowledge Park, 2006).

Harper, P.S., *The Evolution of Medical Genetics: British Perspective* (CRC Press, 2020).

Keller, E.F., *The Century of the Gene* (Harvard University Press, 2000).

Khlinovskaya Rockhill, E., 'On Interdisciplinarity and Models of Knowledge Production', *Social Analysis*, Vol. 51, No. 3, 2007.

Khoury, M.J., 'From Genes to Public Health: The Application of Genetic Technology in Disease Prevention', *American Journal of Public Health*, Vol. 86, No. 12, 1996.

Khoury, M., Little, J. and Burke, W. (eds.), *Human Genome Epidemiology* (Oxford University Press, 2004).

Kroese, M., Zimmern, R., and Sanderson, S., 'Genetic Tests and Their Evaluation: Can we Answer the key Questions', *Genetics in Medicine*, Vol. 6, No. 6, 2004.

Laboratory Services for Genetics: Report of an Expert Working Group to the NHS Executive and the Human Genetics Commission (Department of Health, 2000).

Lenaghan, J., *Brave New NHS? The Impact of the New Genetics on the Health Service* (IPPR, 1998).

Liddell, K. and Hall, A., 'Beyond Bristol and Alder Hey: The Future Regulation of Human Tissue', *Medical Law Review*, Vol. 13, No. 2, 2005.

Louët, S., 'UK Set to Allow Research on Embryonic Cells', *Nature Biotechnology*, Vol. 18, No. 1034, 2000.

Melzer, D., Raven, A., Detmer, D. Ling, T. and Zimmern, R., *My Very Own Medicine: What Must I Know? Information Policy for Pharmacogenetics* (University of Cambridge, 2003).

Meslin, E.M. and Garba, I., 'Biobanking and Public Health: Is a Human Rights Approach the tie that Binds?, *Human Genetics*, Vol. 130, No. 3, 2011.

Mold, A., Clark, P., Millward, G. and Payling, D., *Placing the Public in Public Health in Post-War Britain, 1948-2012* (Palgrave Macmillan, 2019).

Moore, T. and Burton, H., *Genetics Ophthalmology in Focus: A Needs Assessment and Review of Specialist Services for Genetic Eye Disorders* (PHG Foundation, 2008).

Nothing to Hide, Nothing to Fear? Balancing Individual Rights and the Public Interest in the Governance and use of the National DNA Database – A Report by the Human Genetics Commission (Department of Health, 2009).

Ollier, W., Srposen, T. and Peakman, T., 'UK Biobank: From Concept to Reality', *Pharmacogenomics*, Vol. 6, No. 6, 2005.

Our Inheritance, Our Future: Realising the Potential of Genetics in the NHS (Department of Health, 2003).

Parry, B., 'The New Human Tissue Bill: Categorization and Definitional Issues and their Implications', *Genomics, Society and Policy*, Vol. 1, No. 1, 2005.

Ravine, D. and Sampson, J., 'Commentary: The Future Development of Regional Genetics Services will rely on Partnerships', *British Medical Journal*, Vol. 322, 28 April 2001.

Robertson, 'Translating Breakthroughs in Genetics into Biomedical Innovation: The Case of the UK Genetic Knowledge Parks', *Technology Analysis and Strategic Management*, Vol. 19, No. 2, 2007.

Securing Our Future Health: Taking a Long-Term View (HM Treasury, 2002).

Stem Cell Research: Medical Progress with Responsibility (Department of Health, 2000).

Stewart, A., Brice, P., Burton, H., Pharoah, P., Sanderson, S. and Zimmern, R., *Genetics, Health Care and Public Policy: An Introduction to Public Health Genetics* (Cambridge University Press, 2006).

Strathern, M., 'Experiments in Interdisciplinarity', *Social Anthropology*, Vol. 13, No. 1, 2005.

The Advisory and Regulatory Framework for Biotechnology: Report from the Government's Review (Cabinet Office, 1999).

The NHS Plan: A Plan for Investment, A Plan for Reform, Cmnd. 4818-I (London: HSMO, 2000).

Timmins, N., *The Most Expensive Breakfast in History: Revisiting the Wanless Review 20 Years On* (Health Foundation, 2021).

Wallace, S., Lazor, S. and Knoppers, B.M., 'Consent and Population Genomics: The Creation of Generic Tools', *IRB: Ethics and Human Research*, Vol. 31, No. 2, 2009.

Willard, H.F. and Ginsburg, G.S. (eds.), *Genomic and Personalized Medicine: Volume 1 – Principles, Methodology and Translational Approaches* (Academic Press, 2009).

Wonderling, D., Hopwood, P., Cull, A., Douglas, F., Watson, M., Burn, J. and McPherson, K., 'A Descriptive Study of UK Cancer Genetics Services: An Emerging Clinical Response to the New Genetics', *British Journal of Cancer*, Vol. 85, No. 2, 2001.

4

NEW DIRECTIONS

The Bellagio Initiative

In April 2005, 18 delegates from Britain, the United States, Canada, France, and Germany, met at the Rockefeller Centre in Bellagio, Italy. Their number included geneticists, bioethicists, and public health experts. The aim was to agree on a collective definition for the 'blossoming' field of Public Health Genomics.[1] The meeting was initiated by Zimmern's Public Health Genetics Unit with the hope of creating ongoing cooperation between the established centres in Cambridge, Atlanta, and Seattle, while bringing in relevant expertise from other countries. Ron Zimmern, Wylie Burke, and Muin Khoury organised the meeting and invited leaders in their respective fields with whom they were familiar. Delegates included Bartha Knoppers, Theresa Marteau, and Julian Sampson, as well Judith Allanson (Chief of the Department of Genetics at the Children's Hospital of Eastern Ontario), Angela Brand (Professor for Social Medicine and Public Health at the University of Applied Sciences in Bielefeld), Julian Little (Professor of Human Genome Epidemiology at the University of Ottawa), Teri Manolio (Director of Epidemiology and Biometry and the National Heart, Lung and Blood Institute in Bethesda), Tom Murray (President of the Hastings Center in New York), Linda Rosenstock (Dean of the UCLA School of Public Health), and Sian Griffiths (Professor of Public Health at the Chinese University of Hong Kong). Griffiths was also a senior public health physician in Britain. Other organisations such as the Public Health Agency of Canada were also approached. They dispatched a representative in the form of Mohamed Karmali, Director-General of the Laboratory of Foodborne Zoonoses. Karmali had a background in medical microbiology and infectious disease, having been Head of Microbiology at the Hospital for Sick Children in Toronto. As he recalls, 'my job [at Bellagio] was to get the intelligence of what this was all about and filter it back to the Public Health Agency'.

DOI: 10.1201/9781003221760-5

FIGURE 4.1 Delegates at the 2005 Bellagio Meeting. Left to Right: Elena Khilnovskaya Rockhill, Alison Stewart, Julian Little, Hilary Burton, Linda Rosenstock, Tom Murray, Bartha Knoppers, Judith Allanson, Julian Sampson, Muin Khoury, Angela Brand, and Theresa Marteau.

Source: Reproduced with the permission of the PHG Foundation.

Given that this was funded by the Rockefeller Foundation, once guests arrived at its Villa Serbelloni on the shores of Lake Como, they were well taken care of. As Khoury reflects, 'all we had to do was show up, meet for five days, and you know drink wine and eat pasta, have long walks by the lakes'.[2] The free time however was built in around a series of intensive, structured working sessions, which participants found exciting and intellectually stimulating. It was taken as a matter of trust that everyone present was open to thinking issues through from first principles. Such an approach was particularly 'conducive to exchange'.[3] According to Hilary Burton, 'they were very conceptual people … they loved nothing more than bouncing ideas around'.[4]

A set of key questions were put to the participants at the meeting:

- What are the fundamental concepts and scope of public health genetics/ genomics?
- Can the prospects offered by 'personalised medicine' be reconciled with the population-level goals of public health?
- What are the key ethical, legal, and social issues raised by advances in genomics and how can they be addressed most effectively?
- How can we develop a more effective strategy for collecting and evaluating the information on proposed gene–disease associations and gene–environment interactions?
- How can different disciplines work together effectively to achieve shared goals?
- What competencies do health professionals need to enable them to implement the new type of medicine promised by developments in genomics?[5]

At the heart of the discussions was the now well-established proposition that scientists and policymakers increasingly found themselves in the midst of significant change. As the meeting's final report described:

> Modern research in genetics and molecular biology, boosted by information emerging from the Human Genome Project, offers new opportunities for the promotion of population health. Benefits are anticipated through more effective personalised preventive care, disease treatments with better specificity, and innovative drug therapies.[6]

Yet, the delegates agreed that there was still much uncertainty. Genetic developments were likely to be conditional and uneven, and there would be benefits and challenges. Despite genetic advances, most diseases would remain multifactorial in nature – dependent on both genetic and environmental determinants – and not subject to easy solutions. As a result, there was a need to think through a range of practical and philosophical issues and to be as well prepared as possible. The 'Bellagio Group', as they later described themselves, articulated the challenge in these terms:

> Which vision of the future should the prudent clinician believe: A cornucopia of healthcare innovations based on genomics research, or a stream of genetically-based interventions that fail to deliver to the public? We argue that both visions are correct; that genome-based research will offer unprecedented opportunities for improved disease prevention and therapy but will also generate many promising ideas that do not ultimately provide a health benefit … the pressing challenge is to devise an efficient strategy to distinguish innovative advances from false leads.[7]

It was unanimously decided that public health genomics should aim for 'The responsible and effective translation of genome-based knowledge and technologies for the benefit of population health'.[8] The approach was consciously broad – 'genome-based' was chosen as the appropriate term over 'genetics' or 'genomics' so as to encompass all disease-causing gene interactions. Indeed, it was recognised that the field was now best served by using the phrase 'public health genomics' rather than the potentially narrower conception of 'public health genetics'. The underlying aim was that 'genetic determinants should be neither privileged nor unreasonably demonised'.[9] Here, public health genomics was well placed to help bring about a balanced public debate. It could take into account disease prevention at different levels and offer a bridge between individual health and population health, ensuring a solid evidence base for the use of genetic tests, screening programmes, and other interventions. It also had the potential to develop appropriate regulatory and public policy frameworks which considered economic, legal, ethical, and social factors alongside scientific knowledge. It was recognised that this kind of approach had already begun to have an influence on policy debates in Britain.

Ron Zimmern in particular sought to take this dynamic even further and began to conceptualise public health genomics as an 'enterprise'. It was best understood, the meeting's participants agreed, not as a discrete subject or a body of knowledge but as 'a way of working or approaching problems'.[10] The enterprise was encapsulated in a diagram that showed how genome-based science and technology could be translated in order to improve population health. If this translation was to be successfully realised then a 'machinery' would be required.[11] Knowledge generated by the population sciences and the humanities and social sciences, informed by the wider interests of society, would also have to be taken into account. At the heart of the enterprise would be knowledge integration – 'the driving force or engine house'.[12] From this flowed four key areas of activity – informing public policy, developing, and evaluating preventive and clinical health services, communication and stakeholder engagement, and education and training.

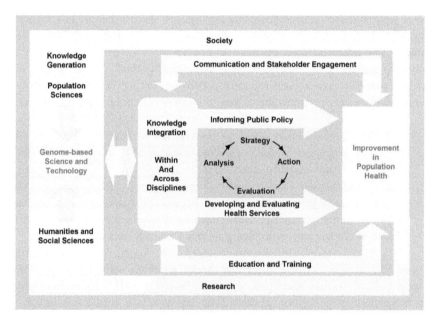

FIGURE 4.2   The Public Health Genomics Enterprise

Source: Reproduced with the permission of the PHG Foundation.

The 'dynamic and interactive nature of the enterprise' was demonstrated by the fluid boundary between the generation of knowledge and its application. The cycle of analysis-strategy-action-evaluation which underpinned the activities was borrowed from established public health practice. Zimmern saw this recognition of public health genomics as a frame of mind or an ongoing enterprise as 'absolutely crucial to my total conceptualisation' of the field.[13] It spoke to what the practice of public health might begin to look in the era of genomic science. It could not consider every element of health and healthcare, but it could become a crucial part of the conversation. To underline this, Zimmern often contrasted this approach with historical conceptions of public health, evoking founding figures such as Edwin Chadwick, in order to demonstrate how much the field had moved on. As he puts it – 'this is what Victorian public health was all about and this is what twenty-first century public health is all about'.[14]

Many of the participants later recognised the Bellagio meeting as a turning point for public health genomics and – to the extent that it was subsequently able to influence it – the development of genetics and genomics policy as a whole. It was an opportunity to formalise much of what had been happening on the ground since that first international meeting, organised by the CDC, in Atlanta in 1998. But it was also a chance to broaden out the field and extend its appeal beyond the small group that had initiated it – to find new 'kindred spirits'.[15] This would not be straightforward, however. The range of interests which could be incorporated was broad and a strategy would be needed to implement this kind of vision. As Karmali reflects:

> On the one hand the multi-disciplinary part, the nature of the participants, contributed to an incredible cross fertilisation of ideas. But on the other hand, the weakness was that it was difficult to know what to do next because there were so many different perspectives and of course everybody was focussed on their perspective on where things should go.[16]

The main initiative to come out of the Bellagio meeting was the establishment of an international network to promote the enterprise of public health genomics. It was given the name GRaPH Int – the Genome-based Research and Population Health International Network. 'Int' was also taken to speak to the fact that the network was integrated and interdisciplinary. There was enthusiasm for a new international organisation, but also many different voices in the room. According to Alison Stewart, 'I don't know how long they spent trying to decide what this thing should be called. Eventually, they came up with GRaPH Int because they wanted public health to be in, they wanted genomics, they wanted international, they wanted population'.[17] The aim would be to consolidate the position of public health genomics and create a critical mass of 'pioneer groups and organisations'.[18] International researchers or those involved in programmes relevant to the field would be able to join, make connections, share knowledge and resources, and develop strategies for working across institutional and disciplinary boundaries. GRaPH Int was also conceptualised as an umbrella organisation or a 'network of networks', as it came to include the US Human Genome Epidemiology Network (HuGENet) and the Network of Investigator Networks in Human Genome Epidemiology, the Canadian Public Population Project in Genomics and Society (P3G Consortium) founded by Knoppers, and the Public Health Genomics European Network (PHGEN), as well as groups such as the PHG Foundation and the CDC Office of Public Health Genomics.[19] Its mission statement established that:

> GRaPH Int is an international collaboration that facilitates the responsible and effective integration of genome-based knowledge and technologies into public policies, programmes, and services for improving population health.[20]

It was the Public Health Agency of Canada that provided the initial funding and took on the role of administrative hub. The feeling at Bellagio had been that it would be better if GRaPH Int was based in a public health service rather than an academic setting. With 'forward thinking' from above at the Public Health Agency, Karmali was able to take advantage of this opportunity to establish an Office of Biotechnology, Genomics and Public Health, which was steadily built up around GRaPH Int between 2005 and 2007.[21] Its aims, drawing on familiar themes were:

> To work with federal and provincial government partners, academia, and other national and international groups, to apply knowledge from advances in biotechnology and genome-based research to prevent disease and improve the health of populations within an ethical, legal, and socially-acceptable framework.[22]

GRaPH Int also had an Executive Group and a Steering Group drawn from the Bellagio meeting participants. It was officially launched in June 2006 at the annual International DNA Sampling Conference on Genomics and Population Health held in Montreal. A first strategic planning meeting was held in Rome in February 2007, at which potential research themes and the ways in which GRaPH Int might be able to 'add value' were discussed. A series of working groups focused on research, education and training, and ELSI issues.[23] Initial objectives were the development of online resources and databases to bring information relevant to public health genomics together, and the creation of internationally agreed educational competencies in genomics.

Under Karmali's leadership, there was also interest in infectious diseases. He was keen to consider the genetic risk in relation to infectious diseases at a population level and begin to be able to assess which groups could be the subject of interventions. There was an expectation that a better understanding of disease mechanisms would lead to initiatives such as predictive genetic screening in order to identify and manage individuals at risk of developing severe, fatal, or chronic outcomes, or becoming persistent carriers of disease pathogens. The wider public health focus inherent to public health genomics and the service organisation setting were important here. As Karmali describes, in contrast to many focussed researchers and academics, 'being a senior bureaucrat allowed you the latitude to think on a higher plane, at a higher level'.[24]

GRaPH Int held its inaugural conference, titled 'Genes for Health', in collaboration with the Human Genetics Society of Australia, in Fremantle in May 2009. The focus was on 'themes at the intersection of diagnostic services, population health and genomics'.[25] Representatives of the PHG Foundation discussed their work around inherited cardiovascular conditions and non-invasive pre-natal diagnosis. Eric Meslin – then Director of the Indiana University Center for Bioethics and a Visiting Professor-at-Large at the University of Western Australia – spoke on the issue of trust and transparency in relation to the development of biobanks. Alongside the work of GRaPH Int at an institutional level, there was still important personal relationship building going on. It was here that Meslin first met

Zimmern. As Meslin recalls, 'Ron was the first, I would say, who 'pitched' this idea [of public health genomics] – a sort of 'What do you think about this? Does it make sense?' You could tell that he already had the basic ideas in mind and was trying to think through all these issues and test them out with people. I could literally peg it to that day, or that coffee break'.[26]

Despite this encouraging start, within the space of a few years GRaPH Int had been wound up. The Canadian funding which had underpinned the Office of Biotechnology, Genomics and Public Health was withdrawn. In 2010, responsibility for the administration and secretariat of GRaPH Int was transferred to the University of Maastricht in the Netherlands, which had recently established an Institute for Public Health Genomics – the first academic centre of its kind in Europe.

Mohamed Karmali was succeeded as Executive Director of GRaPH Int by Angela Brand. Brand had first been a paediatrician with an interest in childhood disorders before training in genetic epidemiology at Johns Hopkins University. She had also worked as a health policy advisor in the United States and Germany, before moving into academia.[27] By the time of the Bellagio meeting, her thinking about the place of genetics in public health had long been complementary to that of Khoury and others.

An important development was the launch of the academic journal *Public Health Genomics* in 2010 with Brand and Knoppers as editors. It was envisaged as the voice of the field and the first journal to focus on 'the responsible and effective translation of genome-based health information and technologies into health interventions and public policies for the benefit of population health'.[28] The journal had originally been established in 1998 as *Community Genetics*, before its publishers decided to adopt a broader and more translational focus and change the name to *Public Health Genomics*.

The editor of *Community Genetics* had been Leo Ten Kate, former Professor of Clinical Genetics at the University of Amsterdam. In the years leading up to the switch, there had been debates about the differences between 'community genetics' and 'public health genomics'. Zimmern's perception for example was that community genetics centred on taking a population approach to Mendelian disorders, whereas public health genomics was a broader conception which considered how genetics played a part in all human disease. It followed therefore that community genetics was a subset of public health genomics. Ten Kate thought that it was the other way round. He defined community genetics 'as a set of activities at the interphase of medical genetics and community medicine aiming to maximize the number of people benefiting from medical genetics while at the same time minimizing potential harm'.[29] The knowledge generated by fields like public health genomics was applied within the context of community genetics he argued, and as such 'Public health genetics is a nuclear family within the extended family of community genetics'.[30]

Ten Kate and colleagues subsequently established a new *Journal of Community Genetics* in 2010 because they felt that the field still had much to offer. While public health genomics and community genetics had much in common there were important differences, they argued: 'The principal aim of public health genetics is to improve population health by reducing disease prevalence. The ultimate aim of community genetics is the well-being of the individual in that population'.[31] Nonetheless, Zimmern maintained that the differences were largely semantic. There was, he said, little to separate the given definition of community genetics – 'the responsible and realistic application of health and disease-related genetic and genomics knowledge and technologies *in human populations and communities to the benefit of individuals therein*' – from the definition of public health genomics – 'the responsible and effective translation of genome-based science and technologies for the benefit of population health'.[32] The debate demonstrates the extent to which there were still translational barriers for public health genomics. 'Public health', 'population health', and 'community' could have subtly different meanings in Britain, the United States, and Europe. There would continue to be differences of emphasis as the implications of genomics emerged.

GRaPH Int's move to Maasctrict was underpinned in part by the perceived success of the Public Health Genomics European Network, which had been launched in 2006 at a conference in Bielefeld, Germany organised by Brand. PHGEN had facilitated important conversations across Europe between those in public health, genetics, bioethics, and law.[33] The Network was funded by the European Commission in two initial phases – PHGEN I from 2006 to 2009 which centred on a mapping exercise and identification of key topics and stakeholders, and PHGEN II from 2009 to 2012 which saw the building of national taskforces in member states. This laid the foundations for future collective approaches around the implementation of genomics in Europe. A declaration was agreed and signed at a key conference in Rome in April 2012. Nonetheless, there was something of a parting of the ways as Brand and European colleagues moved forward with an interpretation of public health genomics that focussed on the more dynamic translation of a wider range of technologies and applications, while the PHG Foundation continued to draw on epidemiological traditions and emphasised the need for a strong evidence base for the implementation of more immediate genomic technologies. Such disagreements about the future direction of public health genomics meant that enthusiasm for GRaPH Int waned somewhat and the initiative was quietly allowed to come to a close.[34]

Even so, GRaPH Int has been seen as a worthwhile – if relatively short term – endeavour. Knoppers suggests that it was able to 'coalesce an idea' and 'served its purpose in bringing the public health genomics community together'.[35] Similarly, Zimmern reflects that 'it was very important for the bonding and getting together a group of ten, fifteen, twenty, thirty people for whom this stuff really did matter'.[36] It was, with hindsight, difficult to move beyond this however and overcome many of the issues that had been identified when GRaPH Int was established. According to Stewart, 'it never completely flew'. Ultimately, GRaPH Int spoke to some of the tensions inherent in public health genomics – 'it was sort of so inclusive that it was easy to just lose sight of what it was'.[37] The initial impetus of bringing disciplines together, developing new ideas, agreeing a way forward and addressing particular problems, was perhaps less apparent in some of the more discursive work which emerged after the Bellagio meeting.[38] Karmali reflects simply that GRaPH Int was 'ahead of its time'.[39]

In parallel with GRaPH Int activities, a follow-up to the Bellagio meeting was held in May 2010 at Ickworth House in Suffolk, with the title 'Public Health in the 21st Century: What Should be the Agenda for Public Health in an era of Genome-Based and Personalised Medicine?' Wylie Burke reflects that there had been much vibrant collaboration during the intervening five years, and participants agreed that a consensus had emerged around three core principles: an 'explicit rejection of genetic exceptionalism', the importance of 'translation supported by a strong evidence base' and an understanding of 'the limits of personalised medicine'. Nonetheless, in the context of rapid developments in genomics and molecular biology, it was felt to be the right time for a 'detailed discussion about how public health should engage with this new scientific agenda' and 'the fundamental issues that might be raised for public health practitioners'.[40] The meeting was felt to be particularly timely in terms of the emerging impact of genomic medicine.[41]

The expanded group of 27 attendees included the core group from the Bellagio meeting, alongside a number of new faces from across a range of different disciplines. The thinking once again was that bringing such a group together in this kind of setting would concentrate minds and allow them to think a number of important issues through. According to Eric Meslin, the Ickworth meeting was 'basically about the future of public health in a genomics world, when the world hadn't even figured out the clinical implications of genomics. It was pretty heady stuff to be thinking about the connection between the exciting work going on in genomics and the contributions to be made by public health'.[42] This was quite jarring he suggests – but 'in a good way'. The main challenges identified related to the wider social and economic contexts in which genomics was operating, health service delivery, and evidence and translation.

There was also now an important discussion about the implications of public health genomics for low and middle-income countries. The vision for public health genomics was given a subtle realignment to include the generation of 'an evidence base to demonstrate when use of genomic information in public health can improve health outcomes in a safe, effective and cost-effective manner'.[43] This would require an appropriate research infrastructure and 'robust multidisciplinary partnerships', including at an international level, underpinned by ethical principles and practices.[44] There was particular interest in a model of 'knowledge synthesis' and translation outlined by Khoury and colleagues.[45] Even so, participants have been somewhat unsure about the real impact of these discussions. While the Ickworth House meeting has been seen as an important step for public health genomics, it has also been suggested that little subsequently changed in practice.

New Independence

Despite the increasing cohesion of those in and around public health genomics, and the increased importance of genetics as a health policy issue in Britain, the fact that the road ahead would not be a smooth one was further demonstrated in 2007 when the funding which had supported the six Genetic Knowledge Parks came to an end. What had started out as a renewable five-year programme had become a finite five-year programme. In 2004, Sally Davies became the Director of Research and Development at the Department of Health. She commissioned an appraisal of funding programmes which concluded in November 2005 that – despite rumours that the number might be reduced from six to one or two – the Genetic Knowledge Parks would not be renewed. Philippa Brice reflects that 'it was a big shock up and down the country when it was all just cut, and I think quite disappointing, because genetics, in five years, was only getting more important'.[46] The overriding sense seemed to be that the vision which had underpinned the creation of the Genetic Knowledge Parks in 2002 had not been realised, and to some extent had itself been unhelpful.

This view was shared by some of those who worked within the Knowledge Parks. Elena Khlinovskaya Rockhill, who conducted an ethnographic study of the Cambridge Genetics Knowledge Park and also attended the Bellagio meeting as a rapporteur, identified a number of common criticisms – 'first, the ambiguity of the initial tender; second, unclear expectations; and, third, 'changing rules' and moving goals, along with an unrealistic time frame, on the part of the funders'.[47] A number of other academic studies have analysed the work of the Genetics Knowledge Parks and described the ways in which their role proved less than straightforward.[48] Relative performance was monitored by the Advisory Group on Genetics Research (AGGR) – comprised of leading scientists and re-presentatives of the pharmaceutical industry and patient groups – with each Knowledge Park required to produce quarterly and annual progress reports. The nature of much of the research being done meant that achievements were difficult to demonstrate within such short timescales. Maxine Robertson suggests this degree of accountability was 'excessive'.[49]

This problem was particularly acute for the Cambridge Genetics Knowledge Park which was distinct from the others with its wider focus on the development of the field of public health genomics. Its remit had been less prescriptive. Staff understood its core function to be the generation of useful knowledge – something which only becomes apparent in the longer term. The kind of networking and collaborating across disciplinary boundaries central to this kind of approach took longer to mature. Links between the research teams – genetic epidemiology, ELSI issues, public health, knowledge and dissemination, and business and administration – remained underdeveloped. This held back the kind of interdisciplinary research that had been envisioned with the appointment of Lecturers in Law, Social Science, Philosophy, and Health Economics. According to Khlinovskaya Rockhill: 'With this absence of the epistemic community, those with an academic background found it necessary to refer back to their disciplinary communities located elsewhere'.[50] There was also less external collaboration than had been expected. According to the authors of the 2010 NIHR-funded analysis *Networks in Health Care: A Comparative Study of Their Management, Impact and Performance*:

There was no reported interaction with important external stakeholders such as NHS managers, commercial companies or NHS commissioners, leading to some concerns in AGGR that the network was the 'odd one out.' Towards the end of the five years, the Department of Health appeared to lose interest.[51]

Robertson even goes as far as to suggest that the competitive ethos of the Knowledge Parks initiative 'severely disrupted' established patterns of collaboration and innovation within the genetics community.[52]

The Department of Health was seen to have lost interest. This reflected an underlying sense of frustration amongst civil servants about apparent differences of outlook between those engaged in applied scientific research and those engaged in research in the social sciences. The life of the policymaker is easier when they are presented with normative solutions to problems rather than new conceptual approaches to understand them. As Mark Bale reflects, once the Genetic Knowledge Parks had been established, there was no way of 'guaranteeing that funding went to the work that we were more interested in, which was the NHS services'.[53] The policy agenda had also moved on since 2002. As with many public policy initiatives, a change of personnel, a change of focus, and the appearance of unanticipated funding constraints caused by the government reassessing priorities can act as a brake on previous ambitions. Dianne Kennard recalls that there were 'a number of our commitments in the White Paper that got caught up in that change … it was less straightforward, shall we say, to follow through on those commitments, because you know, money started to tighten up'.[54] A further factor was that the Genetic Knowledge Parks had been expected to become self-sustaining. Although it was acknowledged that innovative research was being done, there was a feeling that much of the activity fell within the remit of existing sources of funding – particularly the Wellcome Trust which was beginning to invest heavily in genomics and biotechnology – and that the Knowledge Parks did not have a unique enough role to justify continued government support.[55]

Naturally, the Cambridge Genetics Knowledge Park team argued that its remit and function were different and that its wider focus on knowledge generation and policy engagement meant that much of its impact would take place 'upstream' from the applied genetics research. The *Networks in Health Care* study also found that the field of public health genomics – at least in terms of the use of the phrase in textbooks and papers – did grow during this period. More widely it is clear that the Genetic Knowledge Parks helped to raise awareness about genetics and the need to ensure integration between research and practice at an important moment in the story and demonstrated the potential for politicians and policymakers to buy into important ideas – including those inherent to public health genomics – when the right set of circumstances was in place. Yet as Alastair Kent concludes 'the Knowledge Parks were a great idea, but they were too soon … the vision was fine, but the science was not up to the point of delivering it'.[56]

Of the original six, three Knowledge Parks 'survived' in Cardiff, Newcastle, and Manchester. These already had a strong foundation as research centres, were closely aligned with the provision of local genetic services and were able to find other means of support.[57] The Wales Gene Park in Cardiff secured a measure of funding from the Welsh Assembly. The Cambridge Knowledge Park faced a different dilemma. The end of funding from the Department of Health would, without further support, also mean the end of the Public Health Genetics Unit. In 1997, the organisation had been small and able to work with an annual budget of £200,000. From 2002 it had expanded significantly and been working with a budget of around £1 million a year. The PHGU was able to secure external support for a short time, but if it was to continue then a long-term solution would have to be found.

The potential benefits of becoming an independent charitable organisation had been apparent to Zimmern and his team for some time. As the Knowledge Park initiative had demonstrated, the reporting requirements associated with research funding were often onerous. If the PHGU was able to find a different kind of support then it would have more freedom of thought and action. New funding was eventually secured from a private trust which provided a more secure footing.[58] To reflect this change, the PHGU became the Foundation for Genomics and Population Health, and then simply the PHG Foundation. The transition from the Knowledge Park structure and organisation was relatively smooth. Most of the key staff and activities remained the same. According to Carol Lyon, the PHGU's Business Manager:

> I don't think that things changed an awful lot, actually. In the sense that there was still work to be done around building networks, communicating the main issues around public health genomics, raising awareness, we still did comprehensive and very detailed analyses of emerging scientific advances, and what the potential implications were for clinical services.[59]

Brice reflects that 'it was difficult to find suitable funding and set it all up', but that 'the big advantage that it brought was the independence from the Department of Health'.[60]

After the publication of the White Paper *Our Inheritance, Our Future* in 2003, there was a period in which the initiatives it had set in motion – the UK Genetic Testing Network, the National Genetics Education and Development Centre, UK Biobank – were developed and began to operate, without any further significant policy announcements. For the moment, the Department of Health had a clear policy direction. New and transformative work was starting on the development of stem cell therapies but, according to Mark Bale, 'we had been implementing the White Paper ... that money was still being dispersed and the follow-up was still being done'.[61] Although civil servants were aware of it, the wider science of genomics was not yet on the policy radar in a meaningful sense. An implementation review for the White Paper conducted in 2008 painted a positive picture whilst recognising that developments would take place over a long period of time. For example, investment in new technologies and facilities had increased the capacity of genetics testing laboratories, but only 62% of test results were returned in the expected time and to the agreed standards. In order to make 'the best use of available resources' the review agreed to introduce more local flexibility and the ability to prioritise urgent tests.[62]

Other examples of areas where investment was seen to have made an important contribution included a series of pilots around care service pathways and research into new gene therapies. Initiatives such as the UK Genetic Testing Network and the National Genetics Education and Development Centre would have their funding continued, and there was praise for the role of the Human Genetics Commission. However, the Genetics Knowledge Parks were only briefly acknowledged with the recognition that they had 'demonstrated the need for multidisciplinary working in complex areas that have the potential to improve health and quality of life for future generations'.[63] £3 million had also been provided to support the appointment of a new NHS Chair in Pharmacogenetics at the University of Liverpool – the first appointment of its kind. Pharmacogenetics had been identified as a key area for further research, and the position was taken up in 2007 by Munir Pirmohamed.

Fundamentally, the 2008 White Paper progress review recognised that advances in genetics were still accelerating and that if the benefits for health were to be realised they had to be embedded in the NHS. Groups of stakeholders, including organisations representing professionals, specialists, researchers and the NHS, the genetics advisory bodies, and patient groups, argued for continued investment in order to take advantage of new scientific discoveries. Specific areas, including free foetal DNA tests and genetic polymorphisms, were identified, and requests were made for more research in genomics, proteomics (the study of proteins and their interactions and functions in organisms), and transcriptomics (the study of the RNA molecules in a cell) which were important for further understanding the mechanisms that underlie disease and were likely to lead to new and effective treatments.[64] It was during this period that the PHG Foundation produced some of its most impactful work around the role of genetics in mainstream medicine, including reports on ophthalmology and cardiovascular services.[65]

New Strategies

Alongside these policy developments, there were also significant advances in genomic science and the establishment of meaningful provision in genomic medicine. There was no single moment of change – as Peter Harper has described, 'It is neither practical nor helpful to try to separate completely the molecular genetic advances … from those based on the analysis of whole genomes' – but there is a widespread perception of something akin to a 'big bang' taking place during the mid to late 2000s.[66] Those in the field began to make a clearer distinction between clinical genetics and genomics. Underpinned by wider technological developments around high-resolution imaging and computing, whole-genome sequencing became possible and increasingly cheaper, alongside other microbiological techniques such as array CGH and SNP arrays.[67]

In clinical terms, cancer genomics was the first area to substantially develop, and from 2010 whole-genome sequencing was used in the diagnosis of rare diseases and mental disabilities as part of the Deciphering Developmental Disorders study, based at the Wellcome Sanger Institute, with input from each of the 24 Regional Genetics Services in Britain, and funding from the Wellcome Trust and the Department of Health through the Health Innovation Challenge Fund.[68]

In 2006, a new Society for Genomics Policy and Population Health was formed with the aim of reflecting the ongoing development of genomics and the importance of related issues around population health and health policy.[69] It became affiliated with the British Society for Genetic Medicine in 2010, alongside established groups for different clinical specialities. The Society was initially Chaired by Zimmern – with administrative and financial support coming from the PHGU – and then Christine Patch, a leading genetic counsellor, with Alison Hall of the PHGU serving as Secretary. Patch was succeeded in 2008 by Layla Jader, Consultant in Public Health Genomics in Wales. An initially healthy membership of 80–90 included public health physicians, primary care practitioners, lawyers, health economists, and representatives of patients and voluntary organisations with an interest in genetics, as well as a number of clinical geneticists. Subjects considered at meetings and annual conferences included screening programmes, susceptibility testing, understanding of risk, and other ethical issues.

Nonetheless, as with public health genomics as a whole, despite the value of these kinds of activities to those directly involved, it was difficult to measure the wider impact. As Hall reflects, 'The strength was that it was bringing together people from lots of different disciplines to have really rich conversations about issues that everybody was interested in. The weakness was that it didn't have a clear professional identity that mapped onto people's day jobs'.[70] Jader remained frustrated that many of her colleagues in public health for example, despite her best efforts, did not feel the need to get involved with these kinds of initiatives around genetics and genomics.[71] The Society eventually closed in 2013 when its Council could not be assured of securing sustained funding that would leave them with the required degree of independence.[72]

An important recognition of ongoing scientific advances, and one which gave genomics a significant push as a health policy issue, was provided in 2009 by the House of Lords Science and Technology Committee. The Committee was chaired by Lord Sutherland. It established an inquiry into the potential and future needs of genomic medicine, undertaken by a sub-committee chaired by Lord Patel – a former President of the Royal College of Obstetricians and Gynaecologists. Other members of the sub-committee with relevant experience included Baroness O'Neill, Lord Warner, and Lord Winston. The subsequent report *Genomic Medicine* acknowledged the fundamental importance of recent genomic developments and saw them as part of the long history of scientific discovery in Britain:

Every so often, a scientific advance offers new opportunities for making real advances in medical care. From the evidence given to this inquiry, we believe that the sequencing of the human genome, and the knowledge and technological advances that accompanied this landmark achievement, represent such an advance.[73]

Many of the underlying themes discussed in the report were not in themselves new. Indeed, many of the principles that had been established in Our *Inheritance, Our Future* in 2003 were re-stated. For example, it was understood that significant changes would continue for years and that scientific advances would increasingly have medical applications in relation to common complex diseases, not just single-gene disorders. There was a range of ELSI issues associated with these changes that required attention, and there were also commercial opportunities for the bio-technological and pharmaceutical industries which should be taken advantage of.

What had changed between 2003 and 2009 was the nature of the science that underpinned this understanding and the timescales involved. The falling costs of DNA sequencing and the development of new technologies such as genome-wide association studies, following the completion of the Human Genome Project, had opened up new horizons. Genomic advances were already having an impact on mainstream clinical specialties, with new medical applications and means of stratifying individuals in terms of risk and treatment. In addition to single-gene disorders, there was increasing understanding of the role of multiple 'susceptibility genes'. *Our Inheritance, Our Future* had made an important contribution by recognising ongoing changes and the need to prepare the NHS. Now it was necessary to go further.

> The White Paper could hardly have anticipated the remarkable advances since 2003, including the charting of genetic causes of a wide range of common diseases such as diabetes, coronary heart disease and several cancers. These scientific advances are with us now, and the use of genomic diagnostics to provide more rational and increasingly personalised management of common diseases has already started to permeate clinical practice in mainstream specialties across the NHS.[74]

If this was to be achieved then much would depend on the successful translation of genomic research into clinical practice. This had been the subject of an influential report by Sir David Cooksey in 2006, which led to the establishment of the Office for the Strategic Co-ordination of Health Research with the aim of better facilitating translation.[75] Nonetheless, translation remained slow. The 2009 Science and Technology Committee report called for a sustained level of investment in research and a more strategic long-term approach. This was intended as the basis for a further White Paper, alongside the development of a national strategy for the stratified use of medicines – dividing patients into subgroups depending on their therapeutic responses – and preparation for new patterns of provision – including the ways in which tests were commissioned, as genetic treatments continued to impact mainstream medicine.[76] The report also proposed that the utility and validity of new genomic tests should be evaluated by the National Institute for Health and Clinical Excellence (NICE), rather than the UKGTN which was seen to be too small. The size and scope of new challenges highlighted the need for expertise in bioinformatics and the management of genomic databases, as well as modernisation of computing and information technology. The report also called for new national education strategies and core competencies to be agreed upon by the relevant Royal Colleges.

The key point was not so much in the fine detail of the 2009 report, but in the fact that the government, as in 1995 following the House of Commons Science and Technology Committee report on genetics, would be expected to produce a response. This was published in December 2009.[77] As discussed earlier, genomics had not yet become an everyday policy issue and civil servants in the Department of Health were still focused on the implementation of the 2003 White Paper. According to Bale, 'we had to put a lot of work into actually refreshing our memory and beefing up the evidence and explaining what had been done'.[78] This view was shared by Lord Warner who, having helped to launch the White Paper as a Minister in 2003, was aware of the 'enormous expansion of activity' that had taken place in genomics during the intervening years. He felt that the House of Lords Science and Technology Committee 'did a kind of rescue job, in terms of stirring the pot and getting people to take some notice'.[79]

The tone of the government's response was slightly different from that of the Science and Technology Committee report. Though it agreed that genomic medicine would undoubtedly be important, there was less of a sense of urgency about its imminent arrival. The government response accepted few of the Committee's recommendations. Instead, it chose to emphasise key achievements and ongoing plans from the 2003 White Paper. It suggested that the distinction between genomic medicine and ongoing clinical genetics was less clear-cut, so there was no need for a new White Paper. Furthermore, while the wider political context had been important in raising the profile of genetics and bringing substantial investment during the early 2000s, it now arguably held back developments. A general election was on the horizon. As Bale reflects, 'you didn't have to be an expert political Kremlin-watcher to see that there was potentially going to be a change of administration ... we couldn't really commit to doing a White Paper while knowing that that would be on the cards halfway through'.[80] Instead, the government announced that there would be a new 'national strategy' for genomics. According to Bale, this idea stemmed largely from the fact that civil servants did not want to repeat the mistake made in 1995 and allow the government's response to be seen as inadequate:

> We didn't want to have the same criticism of saying 'there is nothing to see here, move along' ... so we actually deliberately took the decision to announce that we would have a strategy ... we need to do this in more

detail, we need to have the luxury of time, we need to wait for the election to happen. What better way than actually getting all the people around the table who gave evidence … and saying, 'What should we do here?'[81]

A Human Genomics Strategy Group (HGSG), made up of key individuals and organisations with insights into genetics research and medicine, was formed by the Department of Health. Its remit was to evaluate the benefits of advances in genomic research and identify barriers to translation.[82] Although the HGSG, and the national strategy which it outlined, would later be seen as effective, there was some initial disappointment at this development. Lord Patel argued that the government's response was 'poor and failed to recognise the reasons for some of our recommendations'.[83]

Ron Zimmern had been invited to give oral evidence to the House of Lords Science and Technology Committee inquiry – a recognition of his unique perspective from the intersection of genomics, policy development and public health. The PHG Foundation also provided written evidence to the inquiry and published an independent response to the report.[84] This was underpinned by a series of workshops with groups of stakeholders – scientists, clinicians, ELSI experts and policymakers – organised with the Cambridge Centre for Science and Policy, whose function was to 'create channels for scientists to consider and communicate the implications of their work for policymakers'.[85] The PHG Foundation's modus operandi was now well established following the Genetics Scenario Project in 2000. Although the 2003 White Paper had been hugely important, the independent response echoed the Science and Technology Committee's suggestion that there was now an 'urgent need' for a national strategy after further rapid technological progress in genomic science.[86] It was agreed that the government had failed to address many of the challenges raised by genomics, but the independent response also sounded a note of caution:

> The House of Lords Committee for Science and Technology should be applauded for recognising that the enormous pace of change in genomic science and technology will require a new strategic phase for implementation in health services. However, we believe that its report on *Genomic Medicine* overestimated the immediate importance of genomics to the prediction and prevention of common diseases, and largely ignored the synergies and opportunities to advance genomic science in the context of the improved diagnosis and treatment of inherited single gene disorders and inherited subsets of complex disease.[87]

The PHG Foundation proposed a more realistic and practical approach that would focus on 'diagnostic and cascade testing for single-gene disorders and inherited subsets of common disease, for which ample evidence of demonstrable health benefits already exists' and only in 'specific instances where utility will be (or has already been) shown for the prediction, diagnosis and management of complex disorders'.[88] As had often been argued, genetic, and now genomic, information should be seen as a complement to other relevant clinical information, not placed above it. A new national strategy would therefore need to focus on translation and implementation in the NHS. A series of recommendations that resonated with the Science and Technology Committee's more practical suggestions and were informed by well-established principles from public health genomics, was put forward. Key actions would include investment in IT and informatics infrastructure, appropriate evaluation mechanisms for tests, ethical handling of genetic information and issues around consent, and the development of bespoke training and education.

The Department of Health established the Human Genomics Strategy Group in 2010, with John Bell as its Chair. Bell was now a well-established leader in the field, having been Regius Professor of Medicine at the University of Oxford since 2002 and Chair of the Office for Strategic Coordination of Health Research since 2008. He was recognised by civil servants as one of those quoted most often in the House of Lords Science and Technology Committee report and able to understand the significance of the genomic changes that were unfolding.[89] Since the publication of key reports by the Genetics Research Advisory Group in 1995, Bell had held on to his vision of a fundamental change in our understanding of disease. In 1998, he prophesied that 'The rapid advances in human molecular genetics seen over the past five years indicate that within the next decade genetic testing will be used widely for predictive testing in healthy people and for diagnosis and management of patients'.[90] In an interview ten years later, Bell reflected that:

> I got the time frame wrong … It's one of my most cited papers because most people hated it. They didn't believe it would happen. I think most people would now accept that the impact of molecular genetics is real – and will become more real. It's very difficult to see how we can continue the same paradigm of healthcare.[91]

Although the initial impact of the House of Lords Science and Technology Committee's report had been limited, there was an opportunity to look again at the possibility of new policy commitments once a general election had been held and the coalition government had taken office in May 2010. Indeed, the push for a genomics strategy aligned with the development of a new life sciences strategy, headlined by Prime Minister David Cameron's announcement of £180 million of investment in December 2011.[92] The vision was for greater collaboration between the NHS and industry, and better use of patient data enabled by genomics and bioinformatics. The Conservative MP George Freeman was one of the main drivers, drawing on his background in venture capital and bioscience in his new role as the government's life sciences advisor from 2011. Freeman saw genomics as a 'disruptive' technology which could empower patients and lay the foundations for innovation and future economic success. This was a message which resonated during a period of tight constraints on public spending. It might also shift the focus of the pharmaceutical industry from drug discovery to drug targeting.[93]

The main result of the HGSG's work was the report *Building on our Inheritance: Genomic Technology in Healthcare*, published in January 2012.[94] Bale suggests that the fact that the report was given a significant launch at the Department of Health, with 'quite firm commitments that we were really going to do something big here' from the Secretary of State for Health, Andrew Lansley, and from Sally Davies, now the Chief Medical Officer, demonstrated the high level of interest. Even in comparison with previous optimistic reports, the scale of the change envisioned was significant. The report said that:

> We are currently on the cusp of a revolution: genomic medicine – patient diagnosis and treatment based on information about a person's entire DNA sequence, or 'genome' – becoming part of mainstream healthcare practice. Increased knowledge and better use of genomic technologies and genetic data will form the basis for a reclassification of disease, with important implications both for predicting natural history and for identifying more effective therapies.[95]

The current applications of genomics were seen as 'just the tip of the iceberg'.[96] Its impact would be felt right across medicine, from primary care to public health, and in diagnosis, treatment, and prevention. The better patient outcomes produced by genomic medicine were 'the ultimate destination of the journey that began with the discovery of the DNA double helix'.[97] *Building on our Inheritance* also reaffirmed that these changes would take place within the NHS, taking advantage of its experience with the development of clinical genetics, and its values, which meant that it was well placed to ensure efficacy, equity, and cost-effectiveness. There would also be significant benefits for the British biomedical industry, enabling it to remain a world leader.

A number of working groups were set up to develop the new national genomics strategy, focussing on Innovation, Service Development, Education, Engagement and Training, and Bioinformatics. A key concern was how to 'mainstream' genomic technology and ensure successful translation. It was recognised that genomic technologies were already helping stratification in cancer treatments and pre-natal screening and that the potential of pharmacogenetics was only just beginning to be realised. *Building on our Inheritance* was also clear that 'there is a strong case that the greatest – and most economically significant – benefits of genomics will be seen in public health'.[98] There was potential for more targeted and less invasive screening, and a better understanding of gene–environment interactions. A better understanding of the causes of common diseases could begin to influence individual behaviours.

In more practical terms, *Building on our Inheritance* called for improved strategic development of services. This would consist of networks of Genomic Technology Centres, Biomedical Diagnostic Hubs and Regional Genetics Centres, brought together as a new kind of 'service delivery infrastructure' encompassing the pathway from initial commissioning through to treatment and counselling for patients and their families. Public health services could also be planned more effectively through 'more comprehensive engagement with genomic technologies from within the public health profession'.[99] Building on existing structures, there would also be an important role for the UKGTN and NICE in developing 'a robust process for the evaluation of clinical validity and utility of all genetic and genomic tests and markers and setting minimum national quality standards'.[100] Bioinformatics would be crucial, with a vision of building a central repository for genetic and genomic data that could easily be accessed. Training in new developments and appropriate safeguards would need to be in place as part of an ongoing consideration of ELSI issues, as well as a mechanism to raise public awareness of genomics and its place in the NHS and wider society. The report also called for a White Paper or 'similar cross-cutting strategic document, which sets out overarching policy direction on genomic technology adoption in the NHS'.[101]

These proposals clearly required significant new investment and capacity. Many of the ideas and perspectives inherent to public health genomics that had been influential in the earlier development of genetics policy, also continued to be important in relation to genomics. Hilary Burton, who had become Director of the PHG Foundation in 2010, was a member of the HGSG. That the work of the PHG Foundation was still in the minds of policymakers was demonstrated by the reaction to its 2011 report *Next Steps in the Sequence*, which has been described as 'critical', 'pioneering' and 'pivotal'.[102] There was certainly a perception that the PHG Foundation was ahead of the game in recognising the importance of faster and cheaper whole-genome sequencing and its implications for clinical practice.

In slightly more measured tones than *Building on our Inheritance*, *Next Steps in the Sequence* suggested that genomics was 'likely to change the current practice of medicine and public health by facilitating more accurate, sophisticated and cost-effective genetic testing'.[103] It was positioned as a contribution to the development of the strategic vision proposed by the House of Lords Science and Technology Committee in 2009, and *Next Steps in the Sequence* was unique in setting out the science behind genomics *and* considering its practical implications. It identified the 'operational barriers' to the adoption and translation of genomics medicine, with a particular 'bottleneck' at the stage of analysis and interpretation of test results. The proposed solution was to simplify the process by only 'answering a specific clinical question' and 'using a process of targeting, prioritising and selecting variants based on an understanding of pathogenicity and frequency'.[104] This was seen as a pragmatic way of integrating whole-genome sequencing into medicine. In the context of the NHS, it was important that implementation happened first where it 'offers clear clinical or cost benefit over existing tests – specifically, for the diagnosis of diseases with a strong heritable component and the management of cancer'.[105] The importance of IT resources, bioinformatics infrastructure, professional training, and appropriate commissioning pathways were also made clear. As might be expected, a pair of scenario workshops with groups of stakeholders – geneticists, oncologists, bioethicists, public health experts, and others – underpinned the report's final recommendations.

Mark Kroese sees *Next Steps in the Sequence* as 'one piece that added to the thinking process' and was valuable for its pragmatic approach, which helped to frame key ideas.[106] According to Brice, *Next Steps in the Sequence* was 'pretty fundamental' and 'fed in quite significantly for the next ten years of policy development'.[107] The launch of the report in October 2011 was attended by a number of Department of Health civil servants and other policymakers, representatives of the NHS, the Wellcome Trust, biotechnological and pharmaceutical companies, health and science policy journalists, and leading clinicians, many of whom – such John Bell and John Burn – were also part of the HGSG. Burton suggests that *Next Steps in the Sequence* was well received because it reflected the ability of the PHG Foundation to have 'credible conversations'.[108] Despite the complexity of the subject, representatives of different disciplines and medical specialties could see how the report was relevant to them. The critical importance of successful knowledge translation was also recognised in other reports such as *Translational Genomics* published by the National Genetics Education and Development Centre in 2013.[109] And many of the themes in *Next Steps in the Sequence* were subsequently developed in the 2014 report *Realising Genomics in Clinical Practice* led by Alison Hall.[110]

100,000 Genomes

After the publication of *Building on Our Inheritance*, an HGSG implementation group was established in 2012 to develop plans for increased laboratory capacity, training, and bioinformatics. Civil servants considered the ways in which these would work within existing NHS structures and in the context of the Department of Health's spending review processes.[111] However, policy was also being made at a higher level, outside of the established structures. According to Bale, 'unbeknownst to me, various advisers had been having other conversations'.[112] John Bell was now the government's official advisor on genetics and had become close to the Prime Minister David Cameron. Bell's conceptualisation of the 100,000 Genomes Project in 2012 was perhaps the most significant development in the recent history of genetics and genomics in Britain. As Sally Davies reflects, 'a lot of the best stuff comes in that way, someone with influence has a really good idea and pushes it'.[113]

The initial opportunity for change had been provided by Cameron's request for a series of briefings on science issues. Bell was one of a group of scientists invited to Downing Street to present their views on the place of genetics and genomics in the NHS and the wider economy, including Mark Walport, Director of the Wellcome Trust, Andrew Wilkie, Nuffield Professor of Pathology at the University of Oxford, and Dame Kay Davies, Professor of Anatomy. Bell describes their surprise at Cameron's commitment to a detailed discussion: 'Bang on 9 o'clock the man himself rocked up, took his jacket off, rolled up his sleeves and said 'OK, I've got three hours. You're going to have to tell me everything you know about genetics'.[114] Cameron had personal experience of NHS genetics counselling through his son Ivan, who had been born with the rare condition

Ohtahara syndrome and died aged six in 2009. It was agreed at the meeting that a major genomics policy initiative would be forthcoming.

A few months later, having been prompted for an update by Cameron, Bell developed the idea of sequencing 50,000 human genomes, in order to understand more clearly how genomics might evolve from a research tool into a core component in everyday clinical medicine. He understood that the cost of whole-genome sequencing was set to fall dramatically – perhaps from as much as $1 million to $1,000 – and that this would permit high frequency, high throughput projects. Bell was also influenced by the 'gamechanger' WGS500 project based at the Wellcome Trust Centre for Human Genetics in Oxford which, in partnership with the biotechnology company Illumina, had sequenced the genomes of 500 patients for whom previous genetic tests had been unhelpful, and identified a number of new causative genes.[115] It was a 'no brainer', Bell suggests, to try something similar at scale within the NHS.[116] A political decision was subsequently taken to increase the headline figure to a more impressive sounding 100,000 genomes.

By September 2012, the key conceptual pieces of the puzzle were in place and the Department of Health was brought in. Bale recalls that 'It was quite a surprise to get a summons from Number Ten. We were doing our, kind of sensible government policy making process' before 'they decided they wanted to announce something big on genomics'.[117] The main details of the project were quickly agreed between Number Ten, Jeremy Hunt, Secretary of State for Health, and David Willetts, Minister of State for Universities and Science. Sufficient funding – initially £100 million – was found. According to Bale, 'There was no money, and then suddenly there was money'.[118]

The 100,000 Genomes Project was officially launched by Cameron in Cambridge in December 2012. He described how, 'By unlocking the power of DNA data, the NHS will lead the global race for better tests, better drugs and above all better care. We are turning an important scientific breakthrough into a potentially life-saving reality for NHS patients across the country'.[119] This was the culmination of years of advances in genomic science, which had been pushed up the policy agenda by the 2009 House of Lords Science and Technology Committee report and the work of the HGCG, but the particular impetus for the 100,000 Genome Project was provided by Bell, Cameron and a small number of others. Bell acknowledges his key role, but as part of a wider effort: 'I had my fingerprints on almost everything that happened', but 'It would be wrong to think that was all me. There were lots of people involved … good smart people making good smart decisions'.[120] Similarly, the high-profile support of the Prime Minister was central to the idea getting off the ground. Genetics and genomics were now a more significant policy issue than ever before. This meant that radical change was possible, but also that the political, economic, and ultimately the clinical stakes were higher, and debates about future directions were likely to be more polarised. For civil servants and other policymakers, this level of interest could be both a positive and negative. According to Bale:

When people ask about the 100,000 Genomes Project – 'what were the best things, what were the worst things' – I say, 'the best thing was having the Prime Minister's backing and the worst thing was having the Prime Minister's backing', because … you felt like you were being constantly observed.[121]

A decision was taken to focus on sequencing the genomes of patients with cancer, rare diseases, and a small number of infectious diseases, to ensure sufficient clinical utility. As CMO, Sally Davies was an important advocate but judged that the NHS, and its commissioning board, would not be able to deliver the project – a decision which George Freeman and others with an eye on the wider industrial strategy appreciated.[122] The solution was to establish a unique 'start-up' – a 'special purpose vehicle' – owned by the Department of Health which would manage contracts for sequencing, data linkage and analysis, and set standards for patient consent.[123] Bell was frustrated by delays but steered the project through the 'politics' of the Department and the NHS.[124]

Genomics England was officially launched in July 2013 to deliver the 100,000 Genomes Project, with funding from the National Institute for Health Research (NIHR), and support from NHS England, Public Health England, and Health Education England. Sir John Chisholm, former Chair of the Medical Research Council, and then head of the defence contractor QinetiQ, was appointed as Executive Chair. Dame Una O'Brien, who was Permanent Secretary at the Department of Health at the time, saw this as an excellent appointment, noting that his 'total genius' was key to developing this novel commercial entity.[125] Sir Mark Caulfield, Professor of Clinical Pharmacology at Queen Mary University, London, who had expertise in the genetics of hypertension, became Chief Scientist. A series of expert groups on data, infrastructure, ethics, and strategic priorities advised the CMO on how best to proceed with the project. The central vision was that the 100,000 Genomes Project would leave the NHS transformed in terms of infrastructure and staff and able to deliver genomics in everyday healthcare. An instruction for the NHS to work with Genomics England came from Ministers, and a series of Genomic Medicine Centres was established. Here patients would be recruited and have samples taken. Caulfield describes these centres as 'an imprimatur of excellence that would allow us to coalesce excellent clinicians, excellent academics, and excellent lab staff around a common infrastructure that would give regional equity of access across England'.[126]

In 2014, a £78 million contract was signed with Illumina to deliver the sequencing. The project was officially completed in 2018. Britain had achieved the first end-to-end whole-genome sequencing 'accredited pipeline' – from NHS labs, to sequencing centres, analysis, and clinics.[127] However, it took time and effort to achieve the necessary degree of coordination and develop the required clinical standards and database capabilities. Frances Flinter, the rare diseases lead for the Genomic Medicine Centre in the South East of England, reflects that 'it was a challenge, and it was very exciting … we had loads of patients without a diagnosis and the idea that we might be able to give them a diagnosis was incredibly exciting'. At the same time, 'it was unbelievably hard work. The political pressure and the pressure from the centre to deliver this project was unlike anything I have ever known before'.[128]

The extent to which some geneticists felt slightly uneasy about the new national approach to genomics is interesting. There was concern that this focus might mean that established clinical genetics, which still had a crucial role to play, would become marginalised. Burton recalls some differences in the HGSG between those like John Bell who were determined about the role of whole-genome sequencing and the arrival of personalised medicine, and those who were somewhat more cautious. Ensuring that the results of whole-genome sequencing were useful and reliable in clinical situations was thought to be more difficult than many optimists realised, and if significant new funding was now available might it not be better spent on improving the genetic testing services that were already available? Bell describes a small number of clinical geneticists as 'technology luddites', who were principally concerned that their comfortable way of life would change.[129] Caulfield reflects that it was hugely important to bring mainstream clinical geneticists on board:

> I remember going to the British Society of Genetic Medicine and a group of clinical geneticists surrounding me and saying, 'This isn't going to work. If you just give us the money we will do exomes [sequence just the protein coding regions of the genome in order to identify variations], we can do it'. I just said to them, 'Look, I understand where you are coming from, why don't you just help me shape this' ... The politicians are expecting this to be done. It is whole genomes, it is not exomes, it is not panels, not any other thing. But I put it to you – I watched you all do incredible things with no resource whatsoever – this is your moment to change it forever. You have all spent your careers working away at this, trying to get the best for families around the country. Let's accept this is a tough challenge and go on it together, and we will empower you to drive it and be at the helm of it, if you do that then we could be looking at something completely different ... if we get it right and we energise this, I can see we have such enthusiasm that we might be able to get this funded longer-term. And fortunately, that is what has happened.[130]

Another indication of the mainstream importance of genomics was Davies' decision to dedicate her 2016 CMO annual report – a longstanding assessment of key issues and the health of the nation – to the topic. This has widely been seen as an important moment. As Eric Meslin suggests, 'there is nothing like a particular leader to draw attention and light'.[131]

Titled *Generation Genome*, the report took stock of how genomics was currently being used in health care and potential for further development.[132] Once again this was understood in terms of advances in technology facilitating better care through improved diagnosis, targeted treatments, and prevention. This was the 'genomics dream'.[133] Cancer and rare diseases were areas that would particularly benefit. The report was optimistic that in the wake of the 100,000 Genome Project: 'we are ahead of the game in transforming our NHS by integrating genomics into the health service in a way other countries dream of'. But Davies also called for further streamlining of genomic laboratories, clearer national standards, better use of data which belied genetic exceptionalism, and better training.[134] She produced the report because 'We were leaders in genomic research but not yet grasping the full opportunities for partners and most people didn't realise what the possibilities of genomic medicine were. It had been a real struggle to set up the 100,000 Genomes Project'.[135] There was a perception amongst some advocates that the NHS could be doing more and in *Generation Genome* Davies described this as a 'perpetual battle'.[136]

Davies had introduced a new approach to annual CMO reports, bringing in a range of expert voices from relevant fields as co-authors. For *Generation Genome,* she commissioned Ron Zimmern to write a chapter on personalised prevention, which he saw as a prestigious opportunity.[137] Pirmohamed wrote a chapter on genomics and therapeutics, which he saw as 'part of the evolution of medicine'.[138] A chapter on the ethical implications of genomic medicine was co-produced by Michael Parker, Jonathan Montgomery, and Anneke Lucassen. Collectively they had significant links with the Human Genetics Commission, the Nuffield Council on Bioethics, UK Biobank, the Human Genomics Strategy Group, the working groups which formed part of the 100,000 Genomes Project, and Genomics England, as well as the PHG Foundation. Lucassen and Parker had also led the influential 'Genethics Club' (now the Genethics Forum) since 2001, which provided an opportunity for researchers and practitioners with an interest in ge-netics to discuss the ethical implications of their work and share best practice.[139] Bale served as Policy Advisor for the report and provided analysis of genomic information and related insurance issues. He sees the reception of *Generation Genome* as 'almost universally positive' and a reflection of a better public under-standing of genomics.[140] The genomic revolution was happening now, and most members of the public were comfortable with the idea.

Where there was controversy it came largely in the form of continuing unease from within the clinical genetics community. In seeking to get her point across in *Generation Genome* Davies described how initially 'genomic services developed as 'cottage industries' built on regional expert presence and local interests and funding'. While this approach had worked well in the past, she argued:

> the scale of the modern NHS and the opportunities offered by genomic medicine meant it is now time to build a first-class genomic service that is scalable, future-proof and delivers value for money'. ... we have never been good at stopping things in the NHS when outdated. This time we must make the changes .... There is a tendency in some parts of the NHS to think of genomics as a thing far in the future, or even worse, a potential burden rather than a boon. While I understand where this view comes from, in the long run it will prevent patients from accessing the best care.[141]

There was, as Bale describes a 'fantastic cadre of people'. However, with 'expanding workloads in clinical genetics, the technology on which they relied was in danger of becoming out of date'.[142] Much of the existing laboratory infrastructure had been built in response to the 2003 White Paper. Alongside their expertise in genetic testing, much of existing expertise in genomic sequencing and interpretation lay with clinical geneticists. As genomics became useful to more clinical specialities, the laboratories in which whole-genome sequencing took place would have to reflect this change. Davies was clear on the need for speed to deliver the government's vision for genomics and the wider life sciences strategy, and there was new money being invested in genomics for research and knowledge integration purposes. But there was no extra funding for existing services.[143] Some individuals found Davies' tone in conveying this message somewhat 'insensitive', especially her repeated referral to clinical genetics as a 'cottage industry'.[144] According to Frances Flinter:

> I think as clinical geneticists, just occasionally, we've felt under threat, because there's been a perception that, within central Government, there isn't necessarily the same degree of understanding as to what it is that we still have to offer. I think there's been a feeling sometimes that … we might become redundant one day, despite the fact that we have huge experience of the interpretation of genomic data in the clinic and are very willing to share this with others'[145]

Others are more understanding of Davies' approach. Zimmern argues that:

> This transformation from genetics to genomics caused the clinical genetics community a lot of angst … Sally was basically cast as the villain by the genetics community in doing what she did, but it was absolutely essential that she did what she did.[146]

A number of influential observers remain philosophical about the impact of these changes. Alastair Kent reflects that:

> From its inception, clinical genetics was very much a kind of individual pursuit … small-scale science, very clever science, but … you were analysing a gene, because that is all the technology allowed you to do … Once you started seeing the potential for high throughput sequencing, then of course the scale of knowledge that was generated increased exponentially, and you could no longer maintain that sort of ideological purity, as it were, because otherwise you would just be left behind.[147]

Similarly, John Burn observes that:

> We are still in the turmoil of breaking up the old system to create the new system … that was one of the reasons why I think the whole genomic

laboratory hub thing was driven through, because there was a clinging to old technologies as a defence against the dark arts … which is a common thing in medicine, generally, that people hang on to technologies for their own sake, and it is a very easy mistake to make.[148]

The steadily evolving nature of clinical practice in response to these changes was demonstrated by the UK Strategy for Rare Diseases published in 2013, which sought to anticipate the ways in which genomic technologies would impact service delivery and ultimately the lives of patients.[149] The PHG Foundation also continued its analytical work, particularly around genetic screening, at a time when new technologies made appropriate appraisal even more important. Emphasis was placed for example on the importance of careful targeting or 'stratification' of particular interventions.[150] Genetic screening was also the subject of a report by the House of Commons Science and Technology Committee in 2014, which argued that more needed to be done to ensure that patients understood both the benefits and the risks.[151] It advocated for a clearer focus on 'informed choice' and for changes to the decision-making processes followed by the National Screening Committee. Burton had briefed the Science and Technology Committee on the complexities around evidence for the potential implementation of new screening programmes, particularly in the light of the launch of the 100,000 Genomes Project and its likely implications and emphasised the importance of keeping ethical issues in mind and keeping screening criteria up to date.

An area which would come to be crucially important was pathogen genomics. Public Health England (the executive agency which led health protection and health improvement initiatives from the centre between 2013 and 2021) launched a Pathogen Genomics Service in 2014, aiming to boost research capacity and facilitate the whole-genome sequencing of disease-causing bacteria and viruses. Particular progress was made around gastrointestinal pathogens.[152] In 2015 the PHG Foundation published the report *Pathogen Genomics Into Practice* which sought to highlight the likely fundamental impact of genomic technology on infectious diseases and set out a 'road map' for successful development and implementation, including significant investment across the health system.[153] The report argued that:

> If through its investment in genomics, England aspires to lead the world in precision medicine, then it must recognise that pathogen genomics, if implemented effectively, represents an opportunity to prove that genomics can truly 'transform' health services. UK scientists and clinicians have laid the foundations for this transformation, but the real challenge begins now with the need for health service leaders to direct and invest to establish the necessary systems and infrastructure to make pathogen genomics part of routine and effective clinical and public health practice.[154]

Mark Kroese reflects that, 'I am very proud. It was an amazing document'.[155] It had become clear to the PHG Foundation that the importance of pathogen genomics was still underappreciated. They worked closely with Sharon Peacock, Professor of Clinical Microbiology at the University of Cambridge, who also recognised this.[156] In 2016 Peacock was the lead author of a chapter on pathogen genomics in *Generation Genome* which also highlighted the changes necessary if whole-genome sequencing was to be adopted in routine public health practice and discussed the potential importance of whole-genome sequencing in overcoming a future pandemic.[157] She had also been a member of the 100,000 Genomes Project working group on strategic priorities and Chaired the subgroup on infectious diseases. As with *Next Steps in the Sequence* in 2012, there was a feeling that *Pathogen Genomics Into Practice* had been timely and well-received. Nonetheless, the subject did not subsequently rise up the agenda and attract significant new investment in the way that had been hoped. Peacock suggests that this was because the importance of pathogen genomic sequencing was not yet widely understood, at least in comparison to more mainstream human genomic sequencing. Many fewer people experienced outbreaks of infectious disease than experienced the impact of cancer, for example.[158]

Given the huge importance that pathogen genomics would come to have from 2019 during the COVID-19 pandemic, and the fact that a comprehensive assessment of the microbiological and bioinformatic capabilities across the country and the infrastructure required to raise its standing had been conducted just a few years before, the report has come to be seen by the PHG Foundation as something of a missed opportunity to build capability.

Unlike with the 100,000 Genomes Project, for example, there was no push to the front of the queue for pathogen genomics. A national pathogen sequencing network, COG-UK – the COVID-19 Genomics UK Consortium, was formed in April 2020 with Peacock as Chair and has played an important role in organising the sequencing of hundreds of thousands of genomes at centres across the country as part of the official COVID-19 response. While COG-UK was brought together very quickly in response to a public health emergency, the scale of pathogen sequencing capacity, and levels of public awareness of the importance of pathogen genomics, have understandably increased significantly since 2020. Thoughts have already begun to turn to how this can be developed into the permanent pathogen sequencing infrastructure that will be needed in future.[159] Peacock suggests that greater equity of access to sequencing and shared data, both at a national and an international level, will be crucial if pathogen genomics is to take on a more established place in everyday public health practice.[160]

In 2017, the Conservative government returned to the idea of an Industrial Strategy as part of its efforts to increase productivity and drive economic growth. As Life Sciences Champion, John Bell was asked to provide further recommendations.[161] Genomics was seen as an area in which significant progress had been made. UK Biobank and the 100,000 Genomes Project were held up as examples of the kind of public–private partnership that would be needed if more investment was going to be forthcoming. This in turn would facilitate more testing, more screening, and further development. Cancer was the area of medicine that would benefit first from the roll-out of whole-genome sequencing. Bell's report also supported calls made in *Generation Genome* in 2016 for a National Genomics Board. Such a body was subsequently established in 2018, chaired by the Health Minister Lord O'Shaughnessy. In introducing a new degree of ministerial direction and bringing together groups of high-profile stakeholders, this has been seen as an important step.[162]

At the heart of subsequent efforts to realise the potential of genomic medicine for routine NHS patients and the life sciences sector has been the Genomic Medicine Service – the first concerted programme of its kind in the world. Launched in 2019, the Genomic Medicine Service has sought to first facilitate faster diagnostics and improved therapies in cancer and rare disease, which will then evolve and reach out into primary and secondary care. The aim of sequencing 500,000 whole genomes by 2024 was set as part of the 2019 NHS Long Term Plan. This was made possible by the introduction of a network of seven genomic laboratory hubs around the country. The UK Genetic Testing Network closed in 2018, with its functions transferred over to the Genomics Unit within NHS England and a new national genomic test directory developed. This process was then taken further with the National Genomics Healthcare Strategy, a ten-year plan outlined in 2020 which, recognising that genomics was increasingly well integrated, will build on existing institutions, funding streams and infrastructure and seek to facilitate better predictive, preventative, and personalised care on the back of whole-genome sequencing. At the same, rationalisation is also likely to mean disruption and loss of some existing expertise.[163]

Notes

1 R. Zimmern and A. Stewart, 'Public Health Genomics: Origins and Concepts', *Italian Journal of Public Health*, Vol. 3, No. 3–2, 2006, p. 9.
2 Interview with Dr Muin Khoury, November 2020.
3 Interview with Professor Bartha Knoppers, March 2021.
4 Interview with Dr Hilary Burton, January 2021.
5 *Genome-Based Research and Population Health*, Report of an expert workshop held at the Rockefeller Foundation Study and Conference Centre, Bellagio, Italy, 14–20 April 2005, p. 5.
6 bid., p. 3.
7 W. Burke, M.J. Khoury, A. Stewart, and R.L. Zimmern for the Bellagio Group, 'The Path from Genome-Based Research to Population Health: Development of an International Public Health Genomics Network', *Genetics in Medicine*, Vol. 8, No. 7, 2006, p. 451.
8 *Genome-Based Research and Population Health*, Report of an expert workshop held at the Rockefeller Foundation Study and Conference Centre, Bellagio, Italy, 14–20 April 2005, p. 7.
9 Ibid., p. 7.
10 Ibid., p. 16.
11 Interview with Dr Ron Zimmern, February 2021.
12 *Genome-Based Research and Population Health*, Report of an expert workshop held at the Rockefeller Foundation Study and Conference Centre, Bellagio, Italy, 14–20 April 2005, p. 16.
13 Interview with Dr Ron Zimmern, February 2021.

14 Ibid.
15 Interview with Dr Muin Khoury, November 2020.
16 Interview with Dr Mohamed Karmali, August 2021.
17 Interview with Alison Stewart, February 2021.
18 R. Zimmern and A. Stewart, 'Public Health Genomics: Origins and Concepts', *Italian Journal of Public Health*, Vol. 3, No. 3–4, 2006, p. 12.
19 P. Brice and R. Zimmern, 'The Public Health Genomics Enterprise' in M. Khoury, S. Bedrosian, M. Gwinn, J. Higgins, J. Ioannidis, and J. Little (eds.), *Human Genome Epidemiology, 2nd Edition: Building the Evidence for Using Genetic Information to Improve Health and Prevent Disease* (Oxford, 2010) p. 54.
20 A. Stewart, M. Karmali, and R. Zimmern, 'GRaPH Int: An International Network for Public Health Genomics' in B.M. Knoppers (ed.), *Genomics and Public Health: Legal and Socio-Ethical Perspectives* (Leiden, 2007) p. 267.
21 Interview with Dr Mohamed Karmali, August 2021.
22 Material provided by Dr Mohamed Karmali.
23 A. Stewart, M. Karmali, and R. Zimmern, 'GRaPH Int: An International Network for Public Health Genomics' in B.M. Knoppers (ed.), *Genomics and Public Health: Legal and Socio-Ethical Perspectives* (Leiden, 2007) p. 276.
24 Interview with Dr Mohamed Karmali, August 2021.
25 P. O'Leary and R.L. Zimmern, 'Genomics and Public Health: Translating Research Into Public Benefit', *Public Health Genomics*, Vol. 13, No. 4, 2010, p. 195.
26 Interview with Dr Eric Meslin, January 2021.
27 Interview with Professor Angela Brand, March 2022.
28 B.M. Knoppers and A.M. Brand, 'From Community Genetics to Public Health Genomics – What's in a Name', *Public Health Genomics*, Vol. 12, No. 1, 2008, pp. 1–3.
29 L.P. Ten Kate, Community Genetics', *International Encyclopaedia of the Social and Behavioural Sciences* (Elsevier, 2001) p. 2359.
30 L.P. Ten Kate, 'Editorial', *Community Genetics*, Vol. 3, No. 1, 2000.
31 J. Schmidtke and L.P. Ten Kate, 'The Journal of Community Genetics', *Journal of Community Genetics*, Vol. 1, No. 1, 2010, p. 1.
32 Emphasis added. R.L. Zimmern, 'A Reply to Community Genetics: 1998–2009 … and Beyond', *Journal of Community Genetics*, Vol. 1, No. 4, 2010, p. 201.
33 Interview with Professor Stefania Boccia, March 2022.
34 Interview with Professor Angela Brand, March 2022.
35 Interview with Professor Bartha Knoppers, March 2021.
36 Interview with Ron Zimmern, February 2021.
37 Interview with Alison Stewart, February 2021.
38 Interview with Dr Hilary Burton, January 2021.
39 Interview with Dr Mohamed Karmali, August 2021.
40 *Public Health in era of Genome-Based and Personalised Medicine* (PHG Foundation, 2010) p. 4.
41 Interview with Dr Ron Zimmern, February 2021.
42 Interview with Dr Eric Meslin, January 2021.
43 *Public Health in era of Genome-Based and Personalised Medicine* (PHG Foundation, 2010) p. 30.
44 Ibid., p. 31.
45 M.J. Khoury, M. Gwinn, and J.P.A. Ioannidis, 'The Emergence of Translational Epidemiology: From Scientific Discovery to Population Health Impact', *American Journal of Epidemiology*, Vol. 172, No. 5, 2010, pp. 517–24.
46 Interview with Dr Philippa Brice, April 2021.
47 E. Khlinovskaya Rockhill, 'On Interdisciplinarity and Models of Knowledge Production', *Social Analysis*, Vol. 51, No. 3, 2007, p. 135.
48 M. Robertson, 'Translating Breakthroughs in Genetics into Biomedical Innovation: The Case of UK Genetic Knowledge Parks', *Technology Analysis and Strategic*

Management, Vol. 19, No. 2, 2007. M. Robertson, 'Experiments in Interdisciplinarity', *Social Anthropology*, Vol. 13, No. 1, 2005. K.G. Hobbs, A.N. Link, and J.T. Scott, 'Science and Technology Parks: An Annotated and Analytical Literature Review', *Journal of Technology Transfer*, Vol. 42, No. 4, 2017.

49 M. Robertson, 'Translating Breakthroughs', p. 201.

50 E. Khlinovskaya Rockhill, 'On Interdisciplinarity', p. 130.

51 E. Ferlie, L. Fitzgerald, G. McGivern, S. Dopson, and M. Exworthy, *Networks in Health Care: A Comparative Study of Their Management, Impact and Performance* (National Institute for Health Research, 2010) p. 64.

52 M. Robertson, 'Translating Breakthroughs', p. 195.

53 Interview with Dr Mark Bale, February 2021.

54 Interview with Dianne Kennard, April 2021

55 Interview with Professor Dame Sally Davies, February 2021.

56 Interview with Alastair Kent, November 2020.

57 Interview with Dr Mark Bale, February 2021.

58 The PHG Foundation's current funding statement is that 'Our work is funded by philanthropic donations, primarily from the Hatton Trust and the WYNG Foundation, along with income from academic grants and collaborations, commercial and public sector commissions and consultancy, and a modest investment portfolio. We are not funded by the University of Cambridge'.

59 Interview with Carol Lyon, April 2021.

60 Interview with Dr Philippa Brice, April 2021.

61 Interview with Dr Mark Bale, February 2021.

62 *Our Inheritance, Our Future: Realising the Potential of Genetics in the NHS – Progress Review* (Department of Health: London, 2008) p. 8.

63 *Our Inheritance, Our Future: Realising the Potential of Genetics in the NHS – Progress Review* (Department of Health: London, 2008) p. 18.

64 Ibid., p. 25.

65 T. Moore and H. Burton, *Genetic Ophthalmology in Focus: A Needs Assessment and Review of Specialist Services for Genetic Eye Disorders* (PHG Foundation, 2008. H. Burton, C. Alberg and A. Stewart, *Heart to Heart: Inherited Cardiovascular Conditions Services – A Needs Assessment and Service Review* (PHG Foundation, 2009).

66 P. Harper, *The Evolution of Medical Genetics: A British Perspective* (CRC Press, 2020) p. 271. Interview with Dr Mark Kroese, April 2021.

67 A.M. Giani, G.R. Gallo, L. Gianfranceschi, and G. Formenti, 'Long Walk to Genomics: History and Current Approaches to Genome Sequencing and Assembly', *Computational and Structural Biotechnology Journal*, Vol. 18, 2020, p. 12.

68 Harper, *The Evolution*, pp. 271–73. H.V. Firth and C.F. Wright, 'The Deciphering Developmental Disorders (DDD) Study', *Developmental Medicine and Child Neurology*, Vol. 53, No. 8, 2011, pp. 702-703.

69 Interview with Alison Hall, May 2021.

70 Ibid.

71 Interview with Dr Layla Jader, June 2021.

72 Ibid.

73 Science and Technology Committee, House of Lords, *Genomic Medicine: Volume 1: Report* (London: The Stationery Office, 2009), p. 5.

74 Ibid.

75 *A Review of UK Health Research Funding* (HM Treasury, 2006).

76 Science and Technology Committee, House of Lords, *Genomic Medicine: Volume 1: Report* (London: The Stationery Office, 2009) p. 39.

77 *Government Response to the House of Lords Science and Technology Committee Inquiry into Genomic Medicine*, Cmnd. 7757, 2009.

78 Interview with Dr Mark Bale, February 2021.

79 Interview with Lord Warner, June 2021.

80 Interview with Dr Mark Bale, February 2021.

81 Ibid.

82 *Government Response to the House of Lords Science and Technology Committee Inquiry into Genomic Medicine*, Cmnd. 7757, 2009, p. 7.

83 Hansard, H.L. Deb., Col. 700, 9 June 2010.

84 *Genomic Medicine: An Independent Response to the House of Lords Science and Technology Committee Report* (PHG Foundation, 2010).

85 Ibid.

86 Ibid., p. 6.

87 Ibid., p. 19.

88 Ibid., p. 6.

89 Interview with Dr Mark Bale, February 2021.

90 J. Bell, 'The New Genetics in Clinical Practice', *British Medical Journal*, Vol. 316, February 14, 1998.

91 G. Watts, 'Professor Sir John Bell, President of the Academy of Medical Sciences', *Clinical Medicine*, Vol. 9, No. 5, 2009, p. 463.

92 *Strategy for UK Life Sciences* (Department for Business, Innovation and Skills, 2011).

93 Interview with George Freeman MP, July 2021.

94 *Building on our Inheritance: Genomic Technology in Healthcare* (Department of Health, 2012).

95 Ibid., p. 14.

96 Ibid., p. 16.

97 Ibid.

98 Ibid.,p. 32.

99 Ibid., p. 38.

100 Ibid., p. 11.

101 Ibid., p. 10.

102 Interview with Dr Mark Kroese, April 2021. Interview with Dr Philippa Brice, April 2021. Interview with Alison Hall, May 2021.

103 *Next Steps in the Sequence: The Implications of Whole Genome Sequencing for Health in the UK* (PHG Foundation, 2011) p. 3.

104 Ibid., p. 4.

105 Ibid., p. 6.

106 Interview with Dr Mark Kroese, April 2021.

107 Interview with Dr Philippa Brice, April 2021.

108 Interview with Dr Hilary Burton, January 2021.

109 *Translational Genomics: The Path from Genomic Insight to Clinical Applications, Licensed Drugs and Treatment Decisions Through Case Examples* (National Genetics and Education and Development Centre, 2013).

110 Interview with Alison Hall, May 2021. *Realising Genomics in Clinical Practice* (PHG Foundation, 2014).

111 Interview with Dr Mark Bale, February 2021.

112 Ibid.

113 Interview with Professor Dame Sally Davies, February 2021.

114 Interview with Professor Sir John Bell, August 2021.

115 J.C. Taylor et al, 'Factors Influencing Success of Clinical Genome Sequencing Across a Broad Spectrum of Disorders', *Nature Genetics*, Vol. 47, No. 7, 2015, pp. 717–26.

116 Interview with Professor Sir John Bell, August 2021.

117 Interview with Dr Mark Bale, February 2021.

118 Ibid.

119 https://www.gov.uk/government/news/dna-tests-to-revolutionise-fight-against-cancer-and-help-100000-nhs-patients.

120 Interview with Professor Sir John Bell, August 2021.

121 Interview with Dr Mark Bale, February 2021.

122 Interview with Professor Dame Sally Davies, February 2021. Interview with George Freeman MP, July 2021.
123 Interview with Professor Sir Mark Caulfield, May 2021.
124 Interview with Professor Sir John Bell, August 2021.
125 Interview with Dame Una O'Brien, February 2022.
126 Interview with Professor Sir Mark Caulfield, May 2021.
127 Ibid.
128 Interview with Professor Frances Flinter, August 2021.
129 Interview with Professor Sir John Bell, August 2021.
130 Interview with Professor Sir Mark Caulfield, May 2021.
131 Interview with Dr Eric Meslin, January 2021.
132 *Annual Report of the Chief Medical Officer 2016: Generation Genome* (Department of Health, 2017).
133 Ibid., p. 4.
134 Ibid.
135 Interview with Professor Dame Sally Davies, February 2021.
136 *Annual Report of the Chief Medical Officer 2016: Generation Genome* (Department of Health, 2017) p. 4.
137 Interview with Dr Ron Zimmern, February 2021.
138 Interview with Professor Sir Munir Pirmohamed, April 2021.
139 Interview with Professor Anneke Lucassen, April 2021.
140 Interview with Dr Mark Bale, February 2021.
141 *Annual Report of the Chief Medical Officer 2016: Generation Genome* (Department of Health, 2017) p. 6.
142 Interview with Dr Mark Bale, February 2021.
143 Interview with Professor Dame Sally Davies, Professor 2021.
144 Interview with Dr Hilary Burton, January 2021.
145 Interview with Professor Frances Flinter, August 2021.
146 Interview with Dr Ron Zimmern, February 2021.
147 Interview with Alastair Kent, November 2020.
148 Interview with Professor Sir John Burn, December 2021.
149 *The UK Strategy for Rare Diseases* (Department of Health, 2013).
150 *Stratified Screening for Cancer: Recommendations and Analysis from the COGS Project* (PHG Foundation, 2014). *Genetic Screening Programmes: An International Review of Assessment Criteria* (PHG Foundation, 2014).
151 House of Commons Science and Technology Committee, *National Health Screening* (London, The Stationery Office, 2014).
152 *Implementing Pathogen Genomics: A Case Study* (Public Health England, 2018).
153 *Pathogen Genomics Into Practice* (PHG Foundation, 2015).
154 Ibid., p. 11.
155 Interview with Dr Mark Kroese, April 2021.
156 Ibid.
157 *Annual Report of the Chief Medical Officer 2016: Generation Genome* (Department of Health, 2017).
158 Interview with Professor Sharon Peacock, February 2022.
159 *Genomics Revolution* (Public Policy Projects, 2021).
160 Interview with Professor Sharon Peacock, February 2022.
161 *Life Sciences Industrial Strategy – A Report to the Government from the Life Sciences Sector* (Office for Life Sciences, 2017).
162 Interview with Dr Mark Bale, February 2021.
163 *Genome UK: The Future of Healthcare* (HM Government, 2020). Interview with Professor Frances Flinter, August 2021.

Bibliography

Annual Report of the Chief Medical Officer 2016: Generation Genome (Department of Health, 2017).

A Review of UK Health Research Funding (HM Treasury, 2006).

Bell, J., 'The New Genetics in Clinical Practice', *British Medical Journal*, Vol. 316, 14 February 1998.

Building on Our Inheritance: Genomic Technology in Healthcare (Department of Health, 2012).

Burke, W., Khoury, M.J., Stewart, A., and Zimmern, R.L.; for the Bellagio Group, 'The Path from Genome-Based Research to Population Health: Development of an International Public Health Genomics Network', *Genetics in Medicine*, Vol. 8, No. 7, 2006.

Burton, H., Alberg, C., and Stewart, A., *Heart to Heart: Inherited Cardiovascular Conditions Services – A Needs Assessment and Service Review* (PHG Foundation, 2009).

Ferlie, E., Fitzgerald, L., McGivern, G., Dopson, S., and Exworthy, M., *Networks in Health Care: A Comparative Study of Their Management, Impact and Performance* (National Institute for Health Research, 2010).

Firth, H.V. and Wright, C.F., 'The Deciphering Developmental Disorders (DDD) Study', *Developmental Medicine and Child Neurology*, Vol. 53, No. 8, 2011.

Genetic Screening Programmes: An International Review of Assessment Criteria (PHG Foundation, 2014).

Genome-Based Research and Population Health, Report of an expert workshop held at the Rockefeller Foundation Study and Conference Centre, Bellagio, Italy, 14–20 April 2005.

Genome UK: The Future of Healthcare (HM Government, 2020).

Genomic Medicine: An Independent Response to the House of Lords Science and Technology Committee Report (PHG Foundation, 2010).

Genomics Revolution (Public Policy Projects, 2021).

Giani, A.M., Gallo, G.R., Gianfranceschi, L., and Formenti, G., 'Long Walk to Genomics: History and Current Approaches to Genome Sequencing and Assembly', *Computational and Structural Biotechnology Journal*, Vol. 18, 2020.

Government Response to the House of Lords Science and Technology Committee Inquiry into Genomic Medicine, Cmnd. 7757, 2009.

Harper, P.S., *The Evolution of Medical Genetics: A British Perspective* (CRC Press, 2020).

Hobbs, K.G., Link, A.N., and Scott, J.T., 'Science and Technology Parks: An Annotated and Analytical Literature Review', *Journal of Technology Transfer*, Vol. 42, No. 4, 2017.

House of Commons Science and Technology Committee, *National Health Screening* (London, The Stationery Office, 2014).

Implementing Pathogen Genomics: A Case Study (Public Health England, 2018).

Khlinovskaya Rockhill, E., 'On Interdisciplinarity and Models of Knowledge Production', *Social Analysis*, Vol. 51, No. 3, 2007.

Khoury, M., Bedrosian, S., Gwinn, M., Higgins, J., Ioannidis, J., and Little, J. (eds.), *Human Genome Epidemiology, 2nd Edition: Building the Evidence for Using Genetic Information to Improve Health and Prevent Disease* (Oxford, 2010).

Khoury, M.J., Gwinn, M., and Ioannidis, J.P.A., 'The Emergence of Translational Epidemiology: From Scientific Discovery to Population Health Impact', *American Journal of Epidemiology*, Vol. 172, No. 5, 2010.

Knoppers, B.M. (ed.), *Genomics and Public Health: Legal and Socio-Ethical Perspectives* (Leiden, 2007).

Knoppers, B.M. and Brand, A.M., 'From Community Genetics to Public Health Genomics – What's in a Name', *Public Health Genomics*, Vol. 12, No. 1, 2008.

Life Sciences Industrial Strategy – A Report to the Government from the Life Sciences Sector (Office for Life Sciences, 2017).

Moore, T. and Burton, H., *Genetic Ophthalmology in Focus: A Needs Assessment and Review of Specialist Services for Genetic Eye Disorders* (PHG Foundation, 2008).

Next Steps in the Sequence: The Implications of Whole Genome Sequencing for Health in the UK (PHG Foundation, 2011).

O'Leary, P. and Zimmern, R.L., 'Genomics and Public Health: Translating Research Into Public Benefit', *Public Health Genomics*, Vol. 13, No. 4, 2010.

Our Inheritance, Our Future: Realising the Potential of Genetics in the NHS – Progress Review (Department of Health: London, 2008).

Pathogen Genomics Into Practice (PHG Foundation, 2015).

Public Health in era of Genome-Based and Personalised Medicine (PHG Foundation, 2010).

Realising Genomics in Clinical Practice (PHG Foundation, 2014).

Robertson, M., 'Experiments in Interdisciplinarity', *Social Anthropology*, Vol. 13, No. 1, 2005.

Robertson, M., 'Translating Breakthroughs in Genetics into Biomedical Innovation: The Case of UK Genetic Knowledge Parks', *Technology Analysis and Strategic Management*, Vol. 19, No. 2, 2007.

Schmidtke, J. and Ten Kate, L.P., 'The Journal of Community Genetics', *Journal of Community Genetics*, Vol. 1, No. 1, 2010.

Science and Technology Committee, House of Lords, *Genomic Medicine: Volume 1: Report* (London: The Stationery Office, 2009).

Stratified Screening for Cancer: Recommendations and Analysis from the COGS Project (PHG Foundation, 2014).

Taylor, J.C. et al., 'Factors Influencing Success of Clinical Genome Sequencing Across a Broad Spectrum of Disorders', *Nature Genetics*, Vol. 47, No. 7, 2015.

Ten Kate, L.P., 'Community Genetics', *International Encyclopaedia of the Social and Behavioural Sciences* (Elsevier, 2001).

Ten Kate, L.P., 'Editorial', *Community Genetics*, Vol. 3, No. 1, 2000.

The UK Strategy for Rare Diseases (Department of Health, 2013).

Translational Genomics: The Path from Genomic Insight to Clinical Applications, Licensed Drugs and Treatment Decisions Through Case Examples (National Genetics and Education and Development Centre, 2013).

Watts, G., 'Professor Sir John Bell, President of the Academy of Medical Sciences', *Clinical Medicine*, Vol. 9, No. 5, 2009.

Zimmern, R.L., 'A Reply to Community Genetics: 1998-2009 … and Beyond', *Journal of Community Genetics*, Vol. 1, No. 4, 2010.

Zimmern, R. and Stewart, A., 'Public Health Genomics: Origins and Concepts', *Italian Journal of Public Health*, Vol. 3, No. 3–2, 2006.

5

CONCLUSION

The aim of this book has been to trace the development of genetics and genomics policy in Britain over the course of the last 40 years. We have seen that genetics has gone from being an area of specialist interest to a relatively small group of researchers and practitioners, to a recognised public policy issue with increasing significance across the fields of medicine and health. Genetics initially emerged as an area of scientific investigation, particularly in terms understanding the mechanism of inheritance and the structure of DNA. As it began to have practical applications in medicine, a specialist field of clinical genetics developed. Services developed around regional centres of expertise. Early leaders were able to shape the place of genetics in research and practice with little central oversight or direction, and where necessary support was lacking, pioneering individuals were able to draw on the support of patient groups and ensure better provision. Though genetics was emerging as a policy issue, the scope of action of civil servants was limited.

As new technologies developed and the clinical implications of genetics increased, a number of professional bodies were able to put forward a more collective position, pushing for the importance of genetics to be more widely recognised and for a greater degree of professionalisation and coordination in the organisation of services. By the 1990s, as genetics increasingly came to be thought of as important in relation to common complex conditions as well as established inherited disorders, it began to appear more frequently on the radar of policy-makers and politicians. Bodies such as the Nuffield Council on Bioethics and the Human Genetics Advisory Commission were established and began to be able to shape the environment in which research and practice took place. A series of reports – especially those commissioned as part of the NHS Research and Development Programme and *Human Genetics: The Science and its Consequences* published by the House of Commons Science and Technology Committee in

DOI: 10.1201/9781003221760-6

1995 – highlighted the impact that genetics was likely to have in future and made the case for greater coordination and oversight.[1]

By the early 2000s, genetics arrived as a clear public policy issue in its own right. A significant expansion of the policymaking infrastructure took place, with a Genetics Policy Unit created inside the Department of Health and bodies like the Human Genetics Commission established. Funding increased and new capacity was built after a significant speech by the Secretary of State for Health, Alan Milburn, in 2001 and the publication of the Genetics White Paper *Our Inheritance, Our Future* in 2003, which helped to put a number of important initiatives, such as the Genetic Knowledge Parks and the UK Genetic Testing Network, in motion.[2] The areas of medicine in which genetics was seen to have a role continued to increase. The completion of the Human Genome Project in 2003 then gave a strong signal of the likely impact of genomics.

After a period of consolidation and implementation, genomics became a subject of great interest following further scientific developments and the beginnings of meaningful provision in genomic medicine. In 2009 the House of Lords Science and Technology published the influential *Genomic Medicine* report, eventually prompting the government into policy action.[3] However, in 2012, outside of the usual policymaking processes, the 100,000 Genomes Project was conceived and given the go-ahead with the aim of directly integrating genomics into the NHS. The progress made has since been given further impetus by the development of the Genomic Medicine Service and the National Genomics Healthcare Strategy. Genetics and genomics now have a well-established place in health policy debates and are widely conceived as being central to the future provision of healthcare services.

In order to understand these changes, we need to take a rounded view. A number of different factors, often in combination, has been important. First, we need to recognise the enduring importance of the scientific and technological advances which have driven genetics forward. Although early breakthroughs in understanding the chromosomal basis of inherited diseases were important, little medical application was possible until sampling techniques were simplified, and diagnostics became easier as part of a gradual shift from the laboratory to the clinic. Important advances in understanding the genetic dimensions of conditions such as thalassaemia and sickle cell anaemia were made during the 1950s and 1960s. From the 1970s, recombinant DNA technology allowed the formation of new sequences, and Sanger sequencing – the first scalable, economic method for reading the whole DNA sequence – was possible, facilitating further genetic discoveries and allowing networks of clinical genetics services to be developed across the country. From the 1980s, the successful mapping of particular disease genes, including Huntington's disease, Duchenne muscular dystrophy, and cystic fibrosis, the development of the polymerase chain reaction method – which allowed small sections of DNA to be amplified for study, and fluorescence in-situ hybridisation – the use of fluorescent probes which stick to specific places on the human chromosome in order to see if they are present, absent, or duplicated, underpinned shifts towards a molecular understanding of disease. As a result, clinical genetics was able to broaden out substantially and begin to be integrated into mainstream medicine.

It was now possible to understand the genetic contribution to an ever-increasing number of inherited disorders, but new technologies also presented the possibility of being able to understand the genetic contribution to common complex conditions. At the same time, the possibilities of genomics started to become apparent. The Human Genome Project was launched in 1989. The race to sequence the whole human genome underpinned further advances such as array CGH testing – through which gains and losses can be ascertained in any part of the genome, and from the late 2000s next-generation sequencing allowed whole genomes to be sequenced cheaply and quickly. Throughout this period there has been a steady drumbeat of technological change, punctuated by moments of significant innovation. Even so, the fact that a new technology exists is not in itself transformative – many of the individuals encountered in the course of this book have wrestled with issues around translation and implementation. The path from the laboratory to the clinic is not usually a straightforward one. But when practical application in a healthcare setting is possible, new technologies become better understood, easier to use, and more widely integrated over time. The cumulative and incremental nature of technological change as a whole is clear, though, an extra push is sometimes required to drive through important developments more quickly.

Second, we need to appreciate the political and administrative context in which these changes were taking place, which at different times have acted as both a catalyst and a brake. As clinical genetics developed as a speciality it naturally came to the attention of civil servants, politicians, and other health policymakers. Genetics slowly began to be thought of as a policy issue. The first Consultant Adviser to the Chief Medical Office on Genetics was appointed in 1972. Nonetheless, it was professional organisations like the Royal College of Physicians and the Clinical Genetics Society which began to establish what an effective clinical genetics service looked like. For the most part it was not thought to be the responsibility of policymakers. There was limited capacity within the Department of Health to take on new responsibilities. When this did begin to change it was the organisation of existing services that was the key issue. In 1993, *Population Needs and Genetic Services: An Outline Guide* was sent out to health service managers and commissioners, alongside a joint 'Professional Letter' from the Chief Medical Officer and the Chief Nursing Officer, to help them better understand genetics and assess provision in their area.[4] However, these were presented as guidelines rather than formal policies. Subsequent calls for a new regulatory body to oversee the development of genetics were resisted for a number of years as Ministers were concerned that issues which had not been their responsibility would become more politicised.

Yet genetics was beginning to appear on the wider political radar as an area of public interest and, following the House of Commons Science and Technology Committee report, the Human Genetics Advisory Committee was eventually formed in 1995.[5] By the early 2000s, a more detailed understanding of genetics had begun to spread from scientific and clinical circles into wider debates about social and economic policy. More joined-up thinking about the development of services emerged, including a recognition that the Department of Health needed to better understand and be well prepared for the clinical changes that genetics would bring. A Genetics Policy Unit was formed and sought to facilitate collaboration and co-ordination and make sure that the NHS was able to realise the potential of genetics. A more formal policy infrastructure developed as a result. The most high-profile example was the Human Genetics Commission, established in 2000 to advise Ministers about priorities in the delivery of services and research. Stories about Dolly the Sheep, BSE (bovine spongiform encephalopathy), and genetically modified foods had coloured popular perceptions of science and such administrative and institutional changes were now seen as a means of addressing these kinds of issues.

Although genetics remained only a relatively small public policy issue as a whole, the 2001 set piece speech by Alan Milburn and the 2003 White Paper demonstrated more clearly than ever before that it could be a mainstream health policy issue, and one in which significant time and money was invested.[6] Milburn himself was more interested in the potential of genetics than any of his predecessors. It was also an issue which could help to frame wider political narratives, particularly the idea that the Labour government was focussing on the future and uniquely interested in the fairness and equity that the NHS was seen to represent. As a result, a number of important policy initiatives were set in motion. When those initiatives began to reach their conclusion without further impetus being provided, it was felt by observers that the Department of Health had, for the moment, started to lose interest. Impetus was subsequently provided by the 2010 House of Lords Science and Technology Committee Report and the emerging reality of genomic medicine.[7] Although the initial impact of the report had been dampened by the expectation of a general election taking place in the near future, once the new coalition government came to office in 2010 there was an opportunity for significant new policy commitments.

The Human Genetics Strategy Group moved the agenda forward, but perhaps the most important abstract driver of policy came from the top down in the interest shown in genetics by the Prime Minister David Cameron. His wish for a major policy initiative on genomic medicine and subsequent support for the 100,000 Genomes Project were critical to their development. We can see therefore that the political and administrative context in which changes were taking place has been a significant factor in the development of genetics and genomics policy. There has been cumulative and incremental change over time in response to developments in genetics research and practice, but there has also been a number of important moments at which a need to address public concerns, or a desire to take advantage of the opportunities that genetics has presented, have provided a political imperative to act, widening the scope of genetics and genomics policy and speeding up its development.

Third, we need to be aware of the economic imperatives which have influenced genetics policy.

Elements of competition and cooperation between scientists and researchers at national and international levels have been important throughout the history of genetics, including moves to ensure that Britain was at the cutting of research into the genetics of blood diseases during the 1960s and 1970s. During the late 1980s, British involvement in the Human Genome Project was presented as being scientifically and economically ambitious. The two reports for the NHS Central Research and Development Committee in 1995, under the leadership of the influential of Sir Michael Peckham, were a product of the then Conservative government's policy agenda around science and technology. The 1993 White Paper *Realising our Potential: A Strategy for Science, Engineering and Technology* had identified genetics as a growth area which could boost the British pharmaceutical and healthcare industries and contribute towards wealth creation.[8] Martin Bobrow's perception that the second report, led by Sir John Bell, was commissioned because Peckham felt that the relatively pragmatic messages in the first report had undersold the potential benefits of genetics, is particularly striking.[9] During the 2000s, policy debates were often underpinned by an understanding that the development of genetics was bound up closely with the commercial success of the biotechnological and pharmaceutical industries and wider economic competitiveness. *Genetics and Health* for example, called for an 'entrepreneurial spirit' and partnerships between industry and academia.[10] This was also part of the thinking behind the Genetics Knowledge Parks. By the 2010s, rapid developments in genomics were also very much seen in the context of the government's wider Life Sciences strategy which aimed to boost economic growth and maintain Britian's leading role.[11]

Taking all of these factors into account, the picture we see is, for the most part, a cumulative or incremental one in which policy issues around genetics and genomics emerged and then gained wider recognition, before policymakers chose to or were required to address them for particular reasons – whether technological, clinical, political, or economic. It seems likely that, once established, the overall direction of travel – specialisation in clinical genetics, moves towards the integration of genetics in mainstream medicine, and the development of genomic medicine – would have been the same across this period without moments of more dramatic change taking place.

However, there are several striking examples of occasions on which the process has been sped up and genetics and genomics have been pushed to the top of the policy agenda. Individual action has often facilitated meaningful change. For example, as one of the leading clinical geneticists in Britain and a close neighbour and advisor to Alan Milburn, Sir John Burn was able to play an important role in getting across the likely importance of genetics in both understandable clinical terms but also political terms during the early 2000s. Sir George Radda, Chief Executive of the MRC, was able to push through the organisation's support for the development of UK Biobank even though it had previously failed to secure funding through more established roots. Sir John Bell was able to conceptualise and successfully make the case for the 100,000 Genomes Project in 2012. This has arguably served to dramatically speed up the process of integrating genomic medicine into the NHS. When change like this occurs it can be valauble, but it can also be disruptive and controversial. Such changes are not without cost. Experience and expertise that has been built up can be lost. Important initiatives can stand or fall on the ability of an individual to influence events, without the need to build a broader consensus, or take a historically grounded long-term view and really plan for successful translation and implementation. If genetics and genomics policy is to be successful in future then it will need to be cohesive and broadly based. It will need to be forward looking and clinically ambitious, whilst also retaining and drawing on knowledge about what has worked well in the past and what is likely to work well in the future.

Alongside this, the book has also sought to understand the role of public health genomics. Although it has been described as a field, a discipline, or a body of knowledge, those in and around public health genomics have most often conceptualised it as an enterprise or a collective way of approaching particular problems. The definition agreed in 2005 was that public health genomics should aim for 'The responsible and effective translation of genome-based knowledge and technologies for the benefit of population health'.[12] We have seen that public health genomics has its origins in historic intersections between the interests of geneticists and public health experts. These manifested themselves in new-born screening programmes for conditions like Phenylketonuria from the 1960s, and then more formally the field of genetic epidemiology, which drew on the longstanding traditions of epidemiology and sought to incorporate new genetic and statistical approaches.

The label 'public health genetics' first emerged in the United States during the 1990s as a key group of public health practitioners, geneticists, and bioethicists, sought to foster more philosophical debates about the potential implications of genetic and genomic medicine, confront difficult ethical and legal questions as well as practical ones, and try to ensure the widest possible population health benefits. The foundational text for public health genetics was arguably Muin Khoury's 1996 article 'From Genes to Public Health' in the *American Journal of Public Health*.[13] In time, Khoury argued, genetic susceptibility could come to be seen as a predictive and modifiable disease risk factor just as much as traditional environmental concerns. Prevention might then follow at a behavioural, environmental, and a clinical level. However, if this promise was to be realised then real thought would have to be given to how genetic advances were translated and applied. Khoury led the CDC Office for Genomics and Public Health from 1997 and made important connections with academics like Wylie Burke at the University of Washington, who was interested in the implications of genetics for routine clinical care. The importance of having a public health workforce which understood and was prepared for the impact of genetics was also widely recognised in the United States.

At the same time, complementary thinking was being done in Britain by Ron Zimmern, Director of Public Health for Cambridge and Huntington Health Authority. With local support he established the Public Health Genetics Unit in Cambridge in 1997, with the aim of getting involved in debates about research, the development of genetic services and genetics policy from the perspective of a non-geneticist and a public health physician. The wider context of public health in Britain was important. Many of the individuals who shaped the field of public health genetics were initially trained in and practiced public health medicine. These experiences often instilled a set of values and a perspective that stayed with them, founded on an appreciation of the value of population-wide approaches to health and healthcare-related questions. This was often married to an interest in the possibilities of genetic and genomic medicine but not subsumed by it.

Public health genetics was informed by a growing recognition of the importance of ELSI issues which emerged alongside the Human Genome Project in the United States. The work of a number of influential international organisations such as UNESCO and the Human Genome Organisation, particularly the ethics committee chaired by Bartha Knoppers, was significant. A declaration on the Human Genome and Human Rights was endorsed by the UN in 1998. There was also a history of critical thinking about ethical issues in Britain. The Public Health Genetics Unit sought to build interest in its work by making connections with leading practitioners and policymakers. In 2000, they provided the most detailed analysis of the impact of genetic medicine to date with the *Genetics and Health* report. This likened the impact of genetics to 'a tidal wave, a tsunami, sweeping all before it as it bursts upon the shore', and highlighted the importance of being able to make informed and systematic judgements about the organisation and funding of health services, setting out proposals for a new national policy framework which was aimed directly at senior policymakers.[14]

By the early 2000s, significant elements of the public health genomics approach were beginning to feed through into policy discussions. Internationally, a series of key connections had been made between Khoury, Zimmern, Burke, Knoppers, and others, during the late 1990s. The importance of these collaborations was demonstrated by the agreement of a shared analytical approach at the meeting in Bellagio, Italy in 2005. It was here that it was decided to use the phrase 'public health genomics' rather than the potentially narrower conception of 'public health genetics'. The approach was consciously broad so as to encompass all disease-causing gene interactions. The underlying aim was that 'genetic determinants should be neither privileged nor unreasonably demonised'.[15]

The consensus at Bellagio was that public health genomics was well placed to help bring about a balanced public debate. It could consider disease prevention at different levels and offer a bridge between individual health and population health, ensuring a solid evidence base for the use of genetic tests, screening programmes, and other interventions, and develop appropriate regulatory and public policy frameworks which considered economic, legal, ethical, and social factors as well as scientific knowledge. The main initiative to come out of the Bellagio meeting was the establishment of the GRaPH Int network to promote the enterprise of public health genomics. Although the initiative was relatively short lived, it has been seen as a worthwhile endeavour. A follow-up meeting was held in May 2010 at Ickworth House in Britain. Having become the PHG Foundation, an independent charitable organisation which did not necessarily rely on research funding,

Zimmern and colleagues were subsequently able to take a broader focus and produce a series of important reports, for example, on the growing place of genetics in mainstream medicine, a detailed independent response to the 2009 House of Lords Science and Technology Committee report, and the influential *Next Steps in the Sequence* report in 2012.[16]

Having established this history, how should we assess the impact and influence of public health genomics? It seems clear that ideas and approaches inherent to public health genomics have often had an important place in the development of genetics and genomics policy in Britain. Their influence might not have always been direct, but it has been appreciable. An example is provided by the 2000 *Genetics and Health* report, of which Zimmern and colleagues at the PHG Foundation remain fond.[17] As he reflects, 'Why I am really proud of this is that we laid out and identified all the issues, and now 25 years later, when we have got a strategy, we haven't missed out on any of the issues'.[18] Although most of the ideas in *Genetics and Health* were not new in themselves, it was the first time that they had been brought together in such a systematic way, and the first time that such breadth of knowledge and expertise – through those scenario planning workshops – had been drawn upon in an attempt to prepare for the changes that genetics would bring.

A number of key issues such as professional education, technological infrastructure, and the coordination of services, did subsequently become the focus of significant policy discussions. The 2003 report *Addressing Genetics, Delivering Health* led by Hilary Burton also highlighted the importance of professional education and called for the establishment of a new Centre for Genetics Education.[19] The idea was picked up in the subsequent White Paper *Our Inheritance, Our Future* and the National Genetics Education and Development Centre was established in Birmingham in 2004.[20] Similarly, over the course of the next few years significant investment in laboratory capacity and other infrastructure followed.

By the 2000s, civil servants in the Department of Health knew that the PHG Foundation could be asked for informal advice, was likely to offer a useful perspective, and could provide experts to sit on relevant advisory groups, working parties and committees, many of which formed part of the emerging policy infrastructure around genetics. The PHG Foundation had an established relationship with the Joint Committee on Medical Genetics, for example, and close links with the UK Genetic Testing Network, playing an important role in the development and application of the gene dossier approach, having introduced the ACCE framework and identified subsequent modifications which took better account of the target population, disease, and the purpose of testing.[21] There was also timely work around the implementation of genetics into mainstream areas of medicine like cardiovascular services and ophthalmology.[22] These were important achievements.

Even so, the influence of public health genomics has often been diffuse and difficult to pin down. There is no handy metric by which its impact can be measured. When reflecting on the place of public health genomics and the work of the PHG Foundation, the one word to which the interviewees for this book kept returning was 'helpful'. With regards to Ron Zimmern as an individual, there can be a little doubt that his role, and the nature of his character, forms an important part of this story. He has been seen as 'brave', as 'a visionary', as someone with 'political nous', and as 'appropriately impatient'.[23] Martin Bobrow reflects that 'I think Ron had an impact because he was Ron and there was no-one else in the trade and because he stuck at it'.[24] Zimmern himself frames his contribution as enabling increasing numbers of practitioners and policymakers to understand that:

> When you are dealing with a new set of technologies, a new set of concepts ... you can't do it just with the science ... You need to amalgamate it with the humanities, the ethics, the law, the philosophy ... you also need the input from epidemiology, management science, information science.

And that:

> There is no human trait, physiological or pathological, that isn't part determined by the genome and part determined by external environmental factors ... you come a cropper if you try to develop any form of science without having regard to these other things.[25]

The fact that this is now more widely understood by practitioners and has come to be an established part of policy debates, particularly in the uniquely British context of implementing genetic and genomic medicine in the NHS, is the most valuable contribution of public health genomics. This is certainly the perception shared by many influential observers. According to Keith Peters:

> What they did was make people who otherwise wouldn't have thought about these issues, think about them. It is not as if the legal and ethical issues, and technological developments in healthcare, are novel, they are not, they happen all the time ... What Ron did, and the [PHG] Foundation did, was spread awareness.[26]

When looking for answers to difficult questions, Peters reflects: 'I think, by and large, they came up with better answers than most people did'.[27] Alastair Kent sees the principal achievement of public health genomics as the building of an 'expectation that you will be able to create a solid evidence-based case for improvement in clinical service delivery, or in public health procedures'.[28] John Burn describes the PHG Foundation as 'a kind of rock ... one of the bricks in the wall, one of the recognised go-to places for getting a kind of measured opinion'.[29] Reflecting on its influence, Sally Davies suggests that:

> It has been immense ... because they got Britain moving and thinking about genomics earlier than most countries ... What would it have been like if we hadn't had [the PHG Foundation] and Ron over the years? We would have got there, but ... we would have been much, much slower.[30]

In 2018 the PHG Foundation became a linked-exempt charity at the University of Cambridge, a move which has been widely seen as a recognition of its achievements over the last 25 years.

FIGURE 5.1 Professor Patrick Maxwell, Regius Professor of Physic at the University of Cambridge, with Dr Ron Zimmern and Dr Mark Kroese of the PHG Foundation in 2018

Source: Reproduced with the permission of the PHG Foundation.

Where there is a little more reticence it is largely because, as many of those in and around the field have recognised, public health genomics has been so inclusive and the range of interests and approaches that can form part of the enterprise has been so broad, that there has been a trade off in terms of cohesion and recognition. Although the label 'public health genetics' can safely be applied to the approach developed by Khoury and others in the United States during the late 1990s and the complimentary thinking done by Zimmern in Britain, and the label 'public health genomics' can be applied to the collective vision set out at the Bellagio meeting in 2005, at some point, especially when considering very recent developments, these labels can cease to be helpful frames of reference. The terms themselves have never been taken up by geneticists, public health practitioners, or policymakers, outside of a relatively small circle. As Mark Kroese, the current Director of the PHG Foundation reflects:

> I don't think it really had a lot of traction in the UK, as a term. As an activity yes, but I don't think anybody more broadly … would identify what has happened in the UK in terms of policy and in terms of implementation, as 'public health genomics' … And I don't think that is a bad thing – because we didn't need it.[31]

The activity and the agenda have been more important than the name.

In reaching these conclusions, this book has provided a history of policy development and process at a high level. This is an important subject and one that has not been considered in detail before. At the same time, there has been much going on beneath the surface that lies outside the immediate remit of this book. Future researchers are likely to find a fertile ground and plentiful material when they consider in more detail issues such as the wider international dimensions of changing public health practice, the rate of integration of genetics and genomics in different countries and healthcare systems, the rate of integration of genetics and genomics in different clinical specialties, public understanding of genetics and the broader social and cultural factors which have influenced its development, the diverse range of patient groups and activists, both lay and medically trained, which have engaged with important debates and helped to facilitate wider changes, and important questions around the sex, gender, age, race, and ethnicity of individuals that have been involved in this story.

Where do all these developments leave us, more than 20 years after Francis Collins gave his Shattuck Lecture to the Annual Meeting of the Massachusetts Medical Society, and more than ten years after his hypothetical patient 'John' was expected to have his whole genome sequenced as part of a routine clinical encounter?[32] Although genomic medicine has moved on significantly and implementation and integration continue, this vision has not been realised. At least on these timescales, there has been an evolution in medicine and health rather than a revolution. Many observers have described the ways in which genetics and genomics remain most important in relation to Mendelian diseases and single-gene subsets of complex disease. There has been a significant impact on cancer genetics, for example, but not yet in relation to other common chronic diseases. Although whole genome sequencing generates large amount of data, only a relatively small proportion can be interpreted and put to effective clinical use. This is likely to change as scientific knowledge improves, but even as part of the 100,000 Genomes Project, many patients have waited longer than initially expected for a meaningful diagnosis or remain undiagnosed.[33]

The discipline and practice of public health also continues to evolve, though the focus often remains on those more immediate and tangible social determinants of health. It seems likely that this will begin to change as it becomes easier to consider genetic determinants alongside other factors, and more practical applications are possible, but it will take time. Patience and openness will be needed. Drawing public health expertise back into conversations about the development and organisation of health services would likely be beneficial and facilitate broader engagement. According to Muin Khoury, 'The application of genomics for population health, is clearly now being evidenced, but I still think we are at an early stage, I don't think we are anywhere near what could … what should happen'.[34] In conceptualising the potential application of genomic knowledge, Khoury and colleagues have come to discuss the idea of 'Precision Public Health', which centres on the use of new forms of data – genomic, environmental, and others – to guide interventions to the right population at the right time.[35] Similarly, in relation to personalised medicine, it seems likely that over time genomics will come to be seen less as a novel technology which might provide answers, and more as one element of achieving better preventative measures and better treatment for individuals in the round. Angela Brand and many European colleagues actually prefer the term 'personal health and care' because it can incorporate new uses of data and thinking outside of established epidemiological patterns.

However, amongst other observers there is still plenty of optimism, and potentially some further hostages to fortune. In 2020 a group of researchers and the US National Human Genome Research Institute set out a '2020 Vision', which included predictions about what genomic medicine would look like by 2030. Among them was the suggestion that 'The regular use of genomic information will have transitioned from boutique to mainstream in all clinical settings, making genomic testing as routine as complete blood counts', and that 'An individual's complete genome sequence along with informative annotations will, if desired, be securely and readily accessible on their smartphone'.[36]

Eric D. Green, the Director of the National Human Genome Research Institute, has argued that:

> The vision for the next phase of human genomics is bolder than ever, and continues to embrace many of the original promises of the Human Genome Project. It does so even as some of the previous predictions about genomic advances are still to be fully realized.
>
> Critics may say such genomics boldness is disingenuous and breeds over-promises, but at every encounter during our recent strategic planning process, we found the optimism and exuberance of our colleagues to be both inspiring and metaphorically intoxicating, pushing us to be even bolder and more willing to take risks in describing where genomics might lead us.[37]

If this is to be the case then, as Green and colleagues recognise, genomics must be guided by established values and principles, successfully navigate the attendant ELSI issues, build sustainable data resources, and ensure that the genomic workforce is adequately trained. While future issues – whether gene editing, artificial intelligence, machine learning, digital technologies, or new diagnostic and therapeutic capabilities that cannot yet be anticipated – may be different to those that have gone before, translation will still need to be effective and responsible, the social and environmental determinants of health will still need to be understood, and genetic determinism will still need to be guarded against. The public will need to have a sound understanding of relative risk and the real meaning of genomic information. Services will need to be equitable, appropriate, and well targeted. It seems likely therefore that, whether we call it 'public health genomics' or something else, many of the issues and approaches discussed in this book will continue to be relevant.

Notes

1 *Genetics Research Advisory Group: A First Report to the NHS Central Research and Development Committee on the New Genetics* (London: Department of Health, 1995). *The Genetics of Common Diseases: A Second Report to the NHS Central Research and Development Committee on the New Genetics* (London: Department of Health, 1995). House of Commons. Science and Technology Committee. *Human Genetics: The Science and Its Consequences* (London: HSMO, 1995).
2 *Our Inheritance, Our Future: Realising the Potential of Genetics in the NHS* (Department of Health, 2003).
3 Science and Technology Committee, House of Lords, *Genomic Medicine: Volume 1: Report* (London: The Stationery Office, 2009).
4 *Population Needs and Genetic Services: An Outline Guide* (Genetics Interest Group, 1993).
5 House of Commons. Science and Technology Committee. *Human Genetics: The Science and Its Consequences* (London: HSMO, 1995).
6 *Our Inheritance, Our Future: Realising the Potential of Genetics in the NHS* (Department of Health, 2003).
7 Science and Technology Committee, House of Lords, *Genomic Medicine: Volume 1: Report* (London: The Stationery Office, 2009).
8 *Realising our Potential: A Strategy for Science, Engineering and Technology*, Cm. 2250, (London: HMSO, 1993).
9 Peter Harper interview with Professor Martin Bobrow, November 2004. *Genetics Research Advisory Group: A First Report to the NHS Central Research and Development Committee on the New Genetics* (London: Department of Health, 1995). *The Genetics of Common Diseases: A Second Report to the NHS Central Research and Development Committee on the New Genetics* (London: Department of Health, 1995).
10 *Genetics and Health: Policy Issues for Genetics Science and Their Implications for Health and Health Service* (Nuffield Trust, 2000) p. 76.
11 *Strategy for UK Life Sciences* (Department for Business, Innovation and Skills, 2011).
12 *Genome-Based Research and Population Health*, Report of an expert workshop held at the Rockefeller Foundation Study and Conference Centre, Bellagio, Italy, 14–20 April 2005, p. 7.
13 M.J. Khoury, 'From Genes to Public Health: The Application of Genetic Technology in Disease Prevention', *American Journal of Public Health*, Vol. 86, No. 12, 1996, pp. 1717–1722.
14 *Genetics and Health: Policy Issues for Genetics Science and Their Implications for Health and Health Service* (Nuffield Trust, 2000) p. 1.
15 *Genome-Based Research and Population Health*, Report of an expert workshop held at the Rockefeller Foundation Study and Conference Centre, Bellagio, Italy, 14–20 April 2005, p. 7.
16 T. Moore and H. Burton, *Genetic Ophthalmology in Focus: A Needs Assessment and Review of Specialist Services for Genetic Eye Disorders* (PHG Foundation, 2008). H. Burton, C. Alberg and A. Stewart, *Heart to Heart: Inherited Cardiovascular Conditions Services – A Needs Assessment and Service Review* (PHG Foundation, 2009). *Genomic Medicine: An Independent Response to the House of Lords Science and Technology Committee Report* (PHG Foundation, 2010). *Next Steps in the Sequence: The Implications of Whole Genome Sequencing for Health in the UK* (PHG Foundation, 2011).
17 *Genetics and Health: Policy Issues for Genetics Science and Their Implications for Health and Health Service* (Nuffield Trust, 2000).
18 Interview with Dr Ron Zimmern, November 2020.
19 H. Burton, *Addressing Genetics, Delivering Health: A Strategy for Advancing Dissemination and Application of Genetics Knowledge Throughout our Health Professions* (PHG Foundation, 2003).
20 *Our Inheritance, Our Future: Realising the Potential of Genetics in the NHS* (Department of Health, 2003).

21 *First Report of the UKGTN: Supporting Genetic Testing in the NHS* (UK Genetic Testing Network, 2007).

22 T. Moore and H. Burton, *Genetic Ophthalmology in Focus: A Needs Assessment and Review of Specialist Services for Genetic Eye Disorders* (PHG Foundation, 2008. H. Burton, C. Alberg and A. Stewart, *Heart to Heart: Inherited Cardiovascular Conditions Services – A Needs Assessment and Service Review* (PHG Foundation, 2009).

23 Interview with Naomi Brecker, February 2021. Interview with Dr Mark Kroese, April 2021. Interview with Alastair Kent, November 2020. Interview with Dr Eric Meslin, January 2021.

24 Interview with Professor Martin Bobrow, November 2020.

25 Interview with Dr Ron Zimmern, November 2020.

26 Interview with Sir Keith Peters, November 2020.

27 Ibid.

28 Interview with Alastair Kent, November 2020.

29 Interview with Professor Sir John Burn, December 2020.

30 Interview with Professor Dame Sally Davies, February 2021.

31 Interview with Dr Mark Kroese, April 2021.

32 F.S. Collins, 'Medical and Societal Consequences of the Human Genome Project', *New England Journal of Medicine*, Vol. 341, No. 1, 1999, pp. 28–37.

33 Interview with Professor Frances Flinter, August 2021. Interview with Professor Anneke Lucassen, April 2021. L.M. Ballard, R.H. Horton. S. Dheensa, A. Fenwick and A.M. Lucassen, 'Exploring Broad Consent in the Context of the 100,000 Genomes Project: A Mixed Methods Study', *European Journal of Human Genetics*, Vol. 28, 2020, p. 732–741.

34 Interview with Dr Muin Khoury, November 2020.

35 M.J. Khoury, M. Engelgau, D.A. Chambers and G.A. Mensah, 'Beyond Public Health Genomics: Can Big Data and Predictive Analytics Deliver Precision Public Health?', *Public Health Genomics*, Vol. 21, No. 5–6, 2018, pp. 244–250.

36 E.D. Green et al., 'Strategic Vision for Improving Human Health at the Forefront of Genomics', *Nature*, Vol. 586, No. 7831, 2020.

37 E.D. Green, 'A Vision for the Next Decade of Human Genomics Research', *Scientific American*, 28 October 2020.

Bibliography

Ballard, L.M., Horton R.H., Dheensa, S., Fenwick, A., and Lucassen, A.M., 'Exploring Broad Consent in the Context of the 100,000 Genomes Project: A Mixed Methods Study', *European Journal of Human Genetics*, Vol. 28, 2020.

Burton, H., *Addressing Genetics, Delivering Health: A Strategy for Advancing Dissemination and Application of Genetics Knowledge Throughout our Health Professions* (PHG Foundation, 2003).

Burton, H., Alberg, C., and Stewart, A., *Heart to Heart: Inherited Cardiovascular Conditions Services – A Needs Assessment and Service Review* (PHG Foundation, 2009).

Collins, F.S., 'Medical and Societal Consequences of the Human Genome Project', *New England Journal of Medicine*, Vol. 341, No. 1, 1999.

First Report of the UKGTN: Supporting Genetic Testing in the NHS (UK Genetic Testing Network, 2007).

Genetics and Health: Policy Issues for Genetics Science and Their Implications for Health and Health Service (Nuffield Trust, 2000).

Genetics Research Advisory Group: A First Report to the NHS Central Research and Development Committee on the New Genetics (London: Department of Health, 1995).

Genome-Based Research and Population Health, Report of an expert workshop held at the Rockefeller Foundation Study and Conference Centre, Bellagio, Italy, 14–20 April 2005.

Genomic Medicine: An Independent Response to the House of Lords Science and Technology Committee Report (PHG Foundation, 2010).

Green, E.D., 'A Vision for the Next Decade of Human Genomics Research', *Scientific American*, 28 October 2020.

Green, E.D. et al., 'Strategic Vision for Improving Human Health at the Forefront of Genomics', *Nature*, Vol. 586, No. 7831, 2020.

House of Commons, 'Science and Technology Committee', *Human Genetics: The Science and Its Consequences* (London: HSMO, 1995).

Khoury, M.J., 'From Genes to Public Health: The Application of Genetic Technology in Disease Prevention', *American Journal of Public Health*, Vol. 86, No. 12, 1996.

Khoury, M.J., Engelgau, M., Chambers, D.A., and Mensah, G.A., 'Beyond Public Health Genomics: Can Big Data and Predictive Analytics Deliver Precision Public Health?', *Public Health Genomics*, Vol. 21, No. 5–6, 2018.

Moore, T. and Burton, H., *Genetic Ophthalmology in Focus: A Needs Assessment and Review of Specialist Services for Genetic Eye Disorders* (PHG Foundation, 2008).

Next Steps in the Sequence: The Implications of Whole Genome Sequencing for Health in the UK (PHG Foundation, 2011).

Our Inheritance, Our Future: Realising the Potential of Genetics in the NHS (Department of Health, 2003).

Population Needs and Genetic Services: An Outline Guide (Genetics Interest Group, 1993).

Realising our Potential: A Strategy for Science, Engineering and Technology, Cm. 2250, (London: HMSO, 1993).

Science and Technology Committee, 'House of Lords', *Genomic Medicine: Volume 1: Report* (London: The Stationery Office, 2009).

Strategy for UK Life Sciences (Department for Business, Innovation and Skills, 2011).

The Genetics of Common Diseases: A Second Report to the NHS Central Research and Development Committee on the New Genetics (London: Department of Health, 1995).

INDEX

Milton Keynes UK
Ingram Content Group UK Ltd.
UKHW022041141024
449569UK00015B/684